Bicycle Design

by Mike Burrows

Edited by Tony Hadland

Company of Cyclists

Title-page.
Author Mike Burrows cornering on an early Windcheetah recumbent tricycle that he designed and built. Yes, the chainset is on the left.

© Mike Burrows, Norwich 2000
Reprinted with minor amendments by the Company of Cyclists, February 2001
Published by Company of Cyclists Publications,
7 Coda Avenue, York YO23 2SE. Tel: 01904 778080
Company of Cyclists edition: ISBN 1-898457-07-7
jm@compofcyclists.demon.co.uk.
www.compofcyclists.com

US edition published by AlpenBooks Press
3616 South Road, C-1, Mukilteo, WA 98275 USA
AlpenBooks Press edition: ISBN 0-9669795-2-4

Our thanks to A. G. Jacobs for his help in making this reprint possible

All rights reserved
Unauthorised duplication
contravenes applicable laws

CONTENTS

- ☐ **Foreword by Richard Ballantine** — 5
- ☐ **Introduction** — 11
- **1. The Engine** — 15
  Basic bio-mechanical information about your body.
- **2. Ergonomics – Sitting Comfortably** — 21
  Riding positions for different disciplines.
- **3. Handling – Positively Round the Bend** — 30
  Steering and staying upright, on and off-road.
- **4. Materials and Processes** — 37
  What different bikes are built from – and why.
- **5. Materials and Form** — 53
  How materials affected the evolution of the bike.
- **6. Aerodynamics** — 63
  How to cheat the wind.
- **7. The Wheel** — 91
  From the cartwheel to the tension spoke wheel – and back again.
- **8. Tyres – Annular Pneumatic Suspension** — 101
  Getting a grip on tarmac and track.
- **9. Transmissions – Gears Without Tears** — 107
  How many, what kind and why the chain is best.
- **10. Suspension – The Good, the Bad and the Bouncy** — 122
  Understanding shock absorbers front and rear.
- **11. Retarding Progress** — 134
  Brakes for racing, touring and trundling around.
- **12. Monoblades and Cantilever Wheels – the Single-Minded Approach** — 137
  Why standing on one leg can make sense.
- **13. Lubrication** — 145
  Ways to keep your bike quiet and efficient.
- **14. Bent Cranks** — 148
  Awful cycling inventions.
- **15. Recommendations – Spending the Money** — 152
  What bike to buy.
- ☐ **Appendix: Reading matter** — 154
  Magazines and books worth reading (or burning).

Eastway, early '80s. The Windcheetah trike is still very much alive, though thankfully furry helmets are a thing of the past.

Foreword
by Richard Ballantine

*M*y most vivid early memory of Mike Burrows is from the Isle of Wight Cycling Festival, Easter 1983, where an embryonic HPV event was dominated by the constant quicksilver movements of a half-dozen low-slung recumbent tricycles, all the more visible because their pilots wore brightly coloured skin-suits and helmets decorated with a furry material in fluorescent hues. The trikes were always on the move, darting and agile, and often, as one would start up, the others would also arc and wheel into motion like a flock of birds. Indeed, the obvious leader of the group, Mike Burrows, with a sharply-defined, aquiline face and bushy crop of jet-black hair framed in electric day-glo plumage, looked exactly like one of the birds of prey he stalks as a bird-watcher – intent, quick, and alert.

The charismatic trikes were the first few of a new machine, the Windcheetah SL 'Speedy', and were clearly a breakthrough in HPV design. The races at the Festival were designed to stimulate the development of street-usable machines, and included a fast downhill with obstacles, as well as tight circuit races. Several competitors crashed on the fierce downhill, unable to corner or brake, or both, but the Speedys streaked through with elan and aplomb. In the circuit races, they were in a class of their own for agility and cornering power.

The Windcheetah SL embodied several of Mike Burrows' characteristics as a cycle designer: a flair for independent thinking; a thorough understanding of engineering; an intuitive sense for the characteristics of materials; and a genius for uniting diverse elements into a package greater than the sum of its parts. Perhaps most importantly, the machine reflected Mike's experience as a rider, and met two objectives not always realised by designers: it was practical, and it worked – 'Speedy' was exactly the right name.

The SL after Windcheetah stands for street legal. The machine was created as trainer for pilots of speed record attempt HPVs, and was designed for daily use. Everything about it was practical, with the width of the track, for example, narrow enough to fit through a standard doorway. Yet it also included Ackerman steering and cantilever wheel axles. Twenty years on, the Windcheetah SL Speedy is still one of the top HPVs in the world. In commercial production by Advanced Vehicle Design, Speedys continue to win races and set records, and riding one is more exhilarating and exciting than ever, because the new models incorporate many refinements. I've never asked beyond this life, but if there is a hereafter, I do surely hope they have a Speedy there.

Born in St. Albans in 1943, Mike providentially was blessed with a father who ran an aeromodelling shop, and from an early age designed and scratch-built model aeroplanes. As is often the case with bright, inquiring minds, Mike found nothing of interest in school, and left with no qualifications at the age of 15 to work in his father's shop. Over the next decade and a bit, he became involved with cars and racing, and worked at various engineering jobs as a machinist.

In 1973, when the price of petrol doubled, Mike dropped cars and purchased his first bike, a Carlton Corsa 5-speed, and took up cycle touring. Ever the builder, in 1976 he constructed a touring bike from Reynolds tubing, using the Corsa as a template. In 1979 he started riding time trials, and in 1980 made his first time trial bike, the start of a commercial career, because there was enough tubing for a second frame, which he sold to a friend. In the same year, Mike attended the first major HPV event in Britain, the Aspro Clear Speed Trials in Brighton. The goal of speed by any means human was the perfect challenge for Mike. By 1981 he had built his first HPV chassis, and 1982/83 saw various refinements culminating in the Windcheetah SL Speedy seen at the Isle of Wright.

The Festival of Cycling also saw the formation of the first HPV association in Europe, the British Human Power Club. Mike, incisive and unhesitating in dealing with technical matters or administrative chaff, yet also gregarious, straightforward, and popular, became the club's founding chairman. He served for a decade or more, over a period when BHPC members produced several record-breaking machines and won many victories in international competitions. Yet it was Mike's work with conventional safety bikes which would have major impact.

The combination of his experiences with aero-modelling, HPVs, and riding time trials, led him to focus on the last great area for improvement in road bikes: aerodynamics. At the time, 1980/81, oval frame tubing and streamlined components were in vogue for road bikes, but more as a marketing fashion than as a true aid to speed. Mike did his own thinking on the subject, and in rapid succession built his first funny bike with a split seat tube and his first aero time trial frame, both in steel. These bikes were advances, but confirmed to him that building bikes from metal sticks could only go so far. Truly efficient aerodynamic designs required new shapes and hence new materials, and in 1982/83 Mike designed and registered a decisive step in the evolution of the bicycle: the first Windcheetah carbon monocoque bicycle.

As I am sure most readers will know, a monocoque is a moulded one-piece frame where chassis and skin are one and the same. A moulded diamond frame has holes, but a proper monocoque is as if deformed from a sphere. The initial Windcheetah Mark I model had conventional forks, but in 1985 on a visit to a cycle museum Mike spotted an 1889 Invincible with mono stays and cantilever wheels. Mike realised this would be a major aerodynamic advantage, and built the Mark Ia model, with a monoblade front wheel strut and cantilever axle. Jim Hendry of the BCF became interested in the bike, and took it to Italy to show to the governing body of international cycle sport, the Union Cycliste International (UCI). In characteristic fashion, in 1986 the UCI banned monocoque frames from sanctioned competition, saying 'The main frame of a bicycle shall be constructed from three tubes of not more than an inch-and-a-half in diameter.'

Mike was now a partner and engineering wizard in a thriving firm making packaging machines for London Transport, BT, and most of the banks, security companies, and fun fairs in England. Approaches to bicycle manufacturers with his new bike designs had proved fruitless, and with the UCI decision, matters could have ended at this stage. However, for Mike the various developmental strands – working in carbon fibre, the function of aerodynamics, and the engineering of features such as cantilever wheels and handlebars – had united in a complete vision, and in 1990 he designed and built the Windcheetah Mark II. The most beautiful machine yet, the Mark II was single-sided, with cantilever wheels front and rear,

Not all bike designers ride every day. Some, however, even do their errands on them…

and a seamless, flowing frame/shell. In 1991 the bike came to the attention of Rudy Thoman of Lotus. Mike and Lotus joined to create a one-off variation of the Windcheetah Monocoque, the LotusSport Pursuit bicycle, and history was made when Chris Boardman rode the machine to a new world record and a Gold Medal at the 1992 Olympics. The fanfare and excitement this caused was considerable, and attention rightly focused as much on the radical new bicycle and its inventor as on the rider.

The next sequence of events was eclectic. Lotus faded out of the cycling picture, and an unknown Scot, Graeme Obree, catapulted to fame by riding a bicycle of unusual pedigree to a new hour record. Unable to afford a high-tech mount, Obree built his own, utilising among other prosaic items, bearings from a washing machine for the bottom bracket. In a logical conjunction, Obree and Burrows met, and Mike introduced Obree to Specialized, who supplied him with wheels. Mike fitted a monoblade to Obree's bike, and also built a replica of the machine, as a spare. Using his own bike with Mike's monoblade, Obree broke the hour record again, and then went his own way. The replica wound up on the Specialized stand at the Interbike show in Las Vegas, where it was spotted by Jan Derkson, ex-roadie then vice-president for Giant Europe. After various negotiations, Mike hired on as a design consultant to the main firm, Giant Bicycles in Taiwan. It seemed an unlikely alliance – a maverick day-glo clad rebel and UCI-gadfly, and the world's largest manufacturer of quality bicycles – but has proved, rather like one of Mike's creations, greater than the sum of its parts.

Giant are a Taiwanese firm, with a Taiwanese culture. Trying to explain what this means is like trying to pounce on a spot that moves farther away the closer you come to it. Although most major bicycle firms have factories in, or source bikes from, Taiwan, firms such as Specialized and Cannondale are native to western bicycle culture and have an ingrained identity and marketing presence. East and West are very different, and despite a history which includes producing several bikes of advanced design and considerable merit, Giant were until recently comparatively unknown in the West. However, their engineering excellence and efficiency in manufacturing bikes was and is second to none. In fact, Giant are a primary source for other manufacturers, and make many of the world's quality bicycles, from top-line racing machines through to easy rider recumbents.

Giant wanted to go one better, and develop bikes which would make people come into the store and say, 'I want a Giant bike'. It is a natural, healthy kind of ambition, and hiring Mike was a strong move in this direction. Mike has the ideas, particularly in new materials, and Giant have the size and expertise to make such ideas happen.

In a global firm such as Giant, a huge number of diverse elements must be unified before new products can be produced and sold. As Mike's sidekick and sometime fly-on-the-wall at Giant gatherings, I watched in utter fascination as Mike evinced a remarkable new quality: a talent for diplomacy. Although Mike is a gentle and generous soul (unless crossed on a race track), he is noted for the vigour of his thoughts and curiosity, and for not mincing words. But at Giant, a virtual United Nations of people from many different countries and cultures, Mike was effectively a multi-lingual translator, not of languages, but of ideas. Mike was able to bridge enormous gaps between people in sales and marketing, and those responsible for manufacturing. As an engineer he understood the problems of product design and engineering very well, and as a child of the Western bicycle renaissance and enthusiastic rider of every kind of cycle from delivery bikes to mountain bikes to HPVs, he had a keen, accurate sense of the marketing side.

The dynamics of a large firm, especially one seriously bent on the objective of building better bicycles, are utterly fascinating. A master frame builder is an artisan and a wonder, but in mass manufacture, an astounding number and variety of things each have to be right in and of themselves, and as well, all smoothly coordinated together. In establishing premier mountain bike and road racing teams and going for the gold in terms of consumer recognition and acceptance, Giant set extremely high goals. Many individuals and processes at Giant were involved in the changes, and as the teething problems settled down and lines of communication opened up, exciting things began to happen. Giant have the size and resources for working with materials which are cheap in bulk production, but expensive to tool up for and capitalise – ideal circumstances for a designer such as Mike, who is inventive yet practical, an essential quality for effective production engineering.

First was the MCR racing bicycle, a full monocoque composite which utilises several of Mike's innovations, most notably wheels with flat

Like the Windcheetah trike, Mike's two-wheeled Ratcatcher features a monoblade front and rear. His latest version, Ratcatcher 9, has a suspension 'fork' and a rear fairing/boot.

composite aero spokes. These are injection moulded carbon-reinforced nylon, and exactly the sort of thing only a firm of Giant's size can economically produce. Another key feature, now available on many Giant bikes, is an adjustable stem. This came about on Mike's first trip to Taiwan, when he saw a design for a new stem on a shopping bike and straight-away proposed producing a much better adjustable model. As Giant would in any case have the cost of re-tooling, Mike argued, it was time to upgrade. Giant followed through on the idea of going better, and Mike's stem, now standard on many bikes, is one that many people in the industry wish they had thought of, too.

The MCR is, barring a new model currently in development, as slippery an aero bike as you can buy. An even more significant racing bicycle, however, was less a question of revolution, than of uniting a number of evolutionary improvements into a cohesive package: the Compact Road Design, or TCR. With a sloping top tube, the TCR frame is light but very stiff, and performance is enhanced with numerous aerodynamic refinements including composite forks, seat post, and spokes. The simplest testimonial to the success of the design is the fact that it is much-copied. Another useful factor of the design is that through the use of adjustable stems and seat posts of varying heights, only three frame sizes are needed to fit 99 per cent of riders. This is a major economy in manufacturing and distribution costs, and the result is that one can purchase a top-line racing bike for an exceedingly reasonable sum.

Some time ago, Mike advanced the idea of a monoblade for mountain bikes, but Cannondale, astute and always ready to advance in technology at any time, were first to the market with their Raven model. Although uptake has been slow, one day all competition mountain bikes will utilise this feature. Meanwhile, Giant have started production of the XTC, a dual-suspension mountain bike which does not bob (i.e. do a pogo-stick act when the bike is honked). Again, the rear suspension design on this bike is evolutionary rather than radical, and Mike modestly claims it took him quite a while to even understand it. Perhaps – but with absolutely no bobbing, the XTC's suspension system is slated to be as much of a benchmark as the first Rock Shox forks.

There are many more goodies in the pipeline, including a sensible lightweight folding bike for motorists, an easy-to-ride SWB recumbent bike, and an ultra-aerodynamic single-sided racing machine which incorporates what will prove to one of Mike's most important inventions, a derailleur unit which operates inside the chainline, and thus can be fully enclosed. The project I'm watching with real interest, though, is one promised by monocoque construction: a lightweight, completely maintenance-free city bike, where the frame encloses all the parts, including the chain, gears, and bottom bracket. Think of it: a bike which is clean, lightweight, and needs no maintenance – that's a combination which, to put it mildly, will open up markets.

It is thrilling that Giant and Mike Burrows are so productive. Developments such as the adjustable stem, the TCR, and the soon-to-be City Bike take place on a scale which can lead and influence millions. Happily, though, some things have not changed. Mike's ever-inquiring mind and bright curiosity means that he loves talking to people – and the benefit is two-way, because Mike has a positive genius for cutting through to essentials. Partly this is due to his practical nature, but also, he truly does love cycling, and wants what he shares and says to be on target. He really talks right with you, and with Tony Hadland's essential and able assistance, he's put it all together into a book which like Mike himself, is straight and to the point, and completely useful. More, the book is a milestone, one that will be enjoyed by anyone seriously interested in bikes, because it is the first to deal properly with the impact of modern materials and manufacturing methods on cycle design.

In a world scaled to producing bikes in seconds and the use of space-age materials, it would be easy for a designer to lose touch. Not Mike. A while ago, Mike rang up to say, 'Well, went out for a mountain bike race, entered the first corner, and woke up on a stretcher'. Tales like this worry me sometimes, but then again, I recently went out to Eastway track with Mike, for the BHPC AGM and to see him race his new Ratcatcher SWB recumbent bike. As I watched Mike jostling for position in the pack, or leaning hard through a 120-degree corner with his eyes intently focused on maintaining a good line, I saw a man both having fun, and very ably working at his profession. Mike clearly has many extraordinary gifts and talents, but the fact that he rides the dirt, track and street with the machines he is building, and finds out for himself what is happening, is what makes him one of the leading bike designers in the world. Long may he ride!

Richard Ballantine

London, 1999

Introduction

The idea behind this book is not exactly new. What I have set out to do is to give the reader a better understanding of:
- that most wonderful of mankind's creations, the bicycle (and some of its relatives),
- what makes it tick; and (for those who are that way inclined),
- how to make it tick faster.

I say it is not a new idea. In fact it is almost as old as the bicycle. In *Bicycles and Tricycles* by Archibald Sharp, first published in 1896, the author complains that, while there are many books on cycling, there is little concerning the design of cycles; and that the periodicals are only concerned with racing and personal matters.

So here I am a century later thinking 'here we go again'. I *would* recommend that you simply buy a copy of Sharp yourself: it is still available and it would save me a lot of writing. *Except* that Sharp goes on to tell his readers they will require a basic knowledge of algebra, geometry and trigonometry. Indeed a large part of the book is given over to the principles of mechanics. The bicycle itself does not appear until page 145!

This work will, I hope, be a bit more accessible. It will certainly contain no reference to algebra, as this is a subject that my pen has not entirely mastered. It will also get rather sooner to its main subject, as I have left out the seemingly obligatory history of the bicycle. That is not what the book is about and it has been done quite well elsewhere. You will, however, find a brief but personal bibliography at the back if you would like to know more.

One area I would like to explore though, if the reader will permit me a bit of a ramble, is the origin of the bicycle. This is not because it has not been written about before, but because it has not been done by the right type of person.

As far as I can see most of the speculation in this area has been done by writers, which is something that I am not. (I hope by the time Tony Hadland has finished the editing and re-writes this will not be too obvious!) I am, however, a 'tinkerer' and maybe even an engineer – which is almost certainly what the Baron Karl von Drais was. For it was he who took that most remarkable first step in the evolution of the bicycle, when he discovered that a vehicle with a pair of in-line wheels does not necessarily do the obvious and fall over. It was a step that, the more I think of it, the more remarkable and significant it has to be. But that is not what comes across when I read the existing histories. It is lumped in with all the other 'steps' – like cranks, chains, etc. – which is to do von Drais a great injustice.

For all the significance of the subsequent innovations, they were all logical and inevitable steps; especially in the contemporary climate of industrial energy. What von Drais did, on the other hand, went way beyond logic or evolution. There was no precedent in nature, no natural forerunner to be improved upon. His running machine was as original as it is possible to be.

Baron von Drais's laufmaschine: the most important bicycle in the world.

Despite having no *natural* forerunners, it had to come from somewhere. There were at the time many four-wheeled vehicles in use, both animal and human-powered. The French célérifère, often misquoted as a bicycle, was one of these. It is usually suggested that one of these devices, built by von Drais, was the starting point for the bicycle. But this is not to see the situation with an engineer's eye. An engineer is a relatively logical person and knows perfectly well what would happen if he took two of the four wheels off his vehicle – it would fall over!

But if what you start with has only one wheel, then your engineer could well see an advantage in adding *another*. The best known one-wheeler of this era was the child's hobby horse. It has been suggested as the starting point by several authors, but I do not myself see our Baron on a child's toy.

I think there were, though, other one-wheel devices that nobody ever mentions – wheelbarrows. For there must have been a great need for cheap specialised transport in Europe at this time for the many artisans and craftspeople. This, I am sure, would have resulted in an enormous variety of handcarts being developed.

Confirmation of something of the sort can be seen in, of all places,

The bicycle has no obvious precursor, but it could be that the wheelbarrow provided von Drais with inspiration.

Taipei airport – a place I visit regularly these days! There you can see a reproduction of a remarkable 16th century wall mural, some 20 metres long, depicting life in a Chinese village. It shows, among other things, people pushing a type of barrow with a large (700 mm?), single, centrally-placed wheel. This is not to suggest that the Chinese invented the bicycle, along with gunpowder, etc. But you can be sure these barrows would have been known about in Europe and adopted, even assuming we did not already use them.

Von Drais had worked for a while in the forestry industry, and at the time was teaching the trade to others. He was also an 'inventor'. Such a man would surely have looked at ways of getting timber out of the forests without using expensive horses. What could be better than a single track vehicle? (Anyone not sure of this should try riding a trike off road!) It seems logical that, having added the second wheel, some form of steering would have been necessary to follow the twists and turns of a forest trail. And having added steering it would not have taken long for an inventor and a group of students to discover the secret of balance without which the bicycle is impossible.

But whichever way it came about, and whether it was genius or luck, it started a chain of events that led us to the modern cycle in all its forms. Seen by many as the 'wallpaper' of the transport world, the bicycle is in fact one of the finest examples of engineering design of all time. It uses so little in the form of material or resources to produce, yet it does so much so efficiently. Cheap healthy transport, enjoyable leisure, exciting sport and no harmful side effects – in fact, the best our little planet has to offer. I just hope I am around in 2017 for the bicentennial party.

The other great name that is always mentioned in connection with the origins of the two-wheeler is Leonardo da Vinci. He has long been one of my heroes, but I have to tell you he had absolutely nothing whatsoever to do with the bicycle.

The reason he is always mentioned is, of course, the famous sketch of *something*. However, no serious historian has ever claimed that the sketch was by da Vinci, and no cycling historian has ever claimed it was a bicycle. The whole thing is, I am afraid, a creation of the media, who have never been keen to allow the truth to spoil a good story.

The original confusion arises, I suspect, from the sketch's superficial similarity to a bicycle. This has caused people to assume that it was at least an elevation or side view of something. I would argue, as an engineer and one who has produced pages of similar doodles (although no paintings of enigmatic ladies as yet), that this is actually a plan view, looking down on something. This being so, you could claim it was a bicycle that had fallen over – but I don't think so.

Even more significantly, though, it is inconceivable that anyone, however gifted (and especially someone who was a draughtsman rather than a tinkerer) would have theorised the bicycle into existence. As I mentioned earlier, there is no natural predecessor for the bicycle, unlike cars that are horseless carriages, aeroplanes that are iron birds, and even the famous helicopter, which has the humble sycamore seed as a logical starting point. This is a case where, for once, necessity was the *daughter* of invention, for we certainly could not do without the bicycle now.

All of which is, of course, pure speculation on my part, as very little history was ever written while it was happening. What does not need to be guessed at, because the facts and figures have been available for a long time (i.e. Sharp), is what affects the performance of this by now 180-year-old device.

However, if you were to go by the information gleaned from the many road tests in our current periodicals, you might imagine the subject was still as mysterious as the dark side of the moon – or involved the anatomy of small furry animals, amphibians, etc. and a cauldron. For on the one hand, they will not question a manufacturer's claim that his bicycle is made from a remarkable new thermoplastic titanium alloy or whatever, despite the obvious nonsense of the statement. Yet at the same time they appear blessed with almost supernatural powers, as they quite regularly inform us that they can feel the difference between various grades of steel tubing, even though there is no instrument yet that can do this!

Actually I should not be too unkind about journalists, a few of whom I can still call friends. I am hoping they will be the first to rush out and buy this book, thus increasing awareness and eliminating the disinformation in one go. And of course, it is a bit unfair to blame the journalists entirely, as there is no current source of information beyond the 'how to adjust your gears' level other than Sharp himself. That is, apart from *Bicycling Science* by Whitt & Wilson, which is the modern equivalent of Sharp. It is even better, but still includes the calculus, etc., and so is not an easy read.

I hope this will be the missing book.

Mike Burrows

PS: This book is written from a largely British perspective. So if you are a reader in a distant land, and some of the things or ideas I cover are not familiar or relevant, I apologise. It is all my brain can cope with!

1. The Engine

*T*his is not an ideal subject for an engineer to tackle. We don't study the human power plant a great deal. And even if we did know how to improve it, I suspect that anything more drastic than a haircut would be frowned on by both authorities and competitors.

But it is a subject that demands some understanding if we are to appreciate why the bicycle has come to be what it is (not to mention why it should definitely *not* be a lot of other things).

So, nothing about diet or training. These will certainly improve our engines and are entirely complementary to improving the cycle itself. But they are *not* engineering. And they are in any case well understood and covered by others.

What you need to know

There are some well-understood factors that fundamentally affect how we generate power, yet which seldom seem to be well explained. This is probably because they do not affect training and related matters. But they do need to be understood by anyone hoping to make a better bicycle.

I started by saying that the human body was not really a proper subject for an engineer. It does, however, have quite a bit in common with the sort of engines we would normally work with. And especially with that bane of our lives, the internal combustion engine. The human body and the internal combustion engine both:
- consume fuel, either in the form of gasoline or sticky buns,
- rely on oxygen to 'burn' the fuel and convert it into useful energy and heat,
- work at remarkably similar levels of efficiency, with some 20-30% ending up propelling our respective vehicles.

Internal combustion engines need to be running nearly flat out to achieve these efficiency figures; we clever apes still do quite well even on tick-over. But whereas our development is now effectively static, our mechanical toys are still in their infancy. They will eventually reach far higher levels of efficiency, making it more difficult for succeeding generations to rationalise the joys of cycling!

Muscle tissue and how it works

Back to our own little engines. We'll ignore for the moment the complex way the body processes its fuel, and go straight to where it ends up: the muscles. Understanding how they work is vital.

Our muscles are made up of bundles of individual fibres that can do one thing and one thing only. They can shorten momentarily when signalled to do so by the brain, causing a leg or arm to move. But the individual fibres cannot sustain this contraction. So, when a sustained force is required, they fire in sequence. That is why we can punch harder than we can push.

Muscle tissue comes in two types, 'white' and 'dark'. The white produces most of our short term sprinting or 'anaerobic' energy. This mostly uses energy stored within the fibres themselves. Such short term power is generally only used for seconds, but can be sustained for a few minutes.

The dark fibres are 'aerobic' – the ones we use for continuous effort at whatever level. They do not store their own energy but rely on the blood to constantly replenish the chemicals they use. The blood in turn is processed by the heart, lungs, etc., to enable it to provide this constant source of energy.

That some of us are good at either short or long distance events is largely determined by the ratio of white to dark tissue in our muscles. This ratio can be influenced by training but was initially set by our genetics. That is why training too specifically for one discipline can adversely affect your performance at the other end of the spectrum.

Both types of tissue work best when used in a cyclic manner. That is to say, contracted and then allowed to recover at short, even intervals, rather than trying to exert a constant pressure. The reason for this is again basic engineering. The blood that is carrying the replacement energy has to flow through the muscle to do its job. This is a lot easier if the muscle is in a relaxed state and all of the small blood vessels are open. To get the blood into a contracted muscle is very difficult. Just imagine trying to pour water down a soft rubber hose that has been stretched tight.

Pedalling cadence

What we need to know to make the most efficient use of our muscles is how often to contract/relax them, and in what ratio. The first bit is easy, as cyclists have been debating it for years. We call it 'cadence' – how fast we pedal. Not only has it been

Optimal pedalling cadence seems to be around 100rpm, though our bodies will tolerate quite a bit either side.

Pedalling speeds

	Distance (miles)	Time	Mph	Gear (inches)	Crank (inches)	Crank speed (rpm)	Foot speed (ft/min)	Est'd power (hp)	Est'd thrust (lbf)
Ordinary, track	¼	30 sec	30	53	5	190	493	1.35	91
	½	72 sec	25	56	5	190	392	1.05	88
		60 min	20.1	59	5½	116	330	0.5	50
Safety, track	⅛	12.4 sec	36.3	90	6¼	136	446	1.6	120
	⅛	12.2	37	68	6½	182	619	1.6	85
	¼	29 sec	29.8	64	6¼	170	520	1.3	83
	⅛	11.5 sec	39	90	6½	145	473	1.65	115
Safety, track, motorcycle paced		60 min	40.1	106	6¾	126	445	0.5	37
		60 min	56	139	6½	134	456	0.5	36
		60 min	61.5	144	6½	143	488	0.5	35
		60 min	71	180	6½	133	454	0.5	36
		60 min	76	191	6½	134	454	0.5	36
Train-paced	1	57 sec	62	104	6½	198	670	1.2	59
Road safety bicycle	25	52 min	28.8	90	6⅞	102	370	0.6	54
	100	4 h	25	85	6⅞	99	368	0.5	45
	480	24 h	20	80	6⅞	84	310	0.25	26
	100	4h 28 min	22.4	81	6½	93	316	0.5	52
Road, tourist			10	68	6⅞	50	180	0.09	16
			12	68	6⅞	61	220	0.11	16
			16	75	6⅞	74	266	0.2	24
			18.5	75	6⅞	85	305	0.3	32

Sources: A.C. Davison, Pedalling speeds, Cycling (20 January 1933): 55-56; H. H. England, I call on America's largest cycle maker, Cycling (25 April 1957): 326-327; 'Vandy,' the unbeaten king, Cycling (11 March 1964): 8; Marcel De Leener, Theo's hour record, Cycling (7 March 1970): 28.

much debated over the years but an awful lot of nonsense has been (and still is) talked about it.

When I first came into cycling in the early 1970s the magic number was, I was assured, 100. That is, a hundred revolutions of the cranks per minute. This was what all good cyclists (or at least good time triallists who had traditionally ridden fixed-wheel and were thus able to set this factor quite accurately) should aim for.

Some 25 years and a lot of research later I see no reason to argue with this figure. For a rider of average height and build interested in maximum power output over short to medium distance events it has to be close to optimum. The problem comes when we start talking about track racing. Then, for some reason that has thus far eluded me, everyone gears down despite the obvious lack of hills, little or no wind and somewhat higher speeds.

Certainly over shorter distances, as the power required increases the pedalling rate has to be increased. Our ability to spin seems less limiting than our ability to push. This makes sense bearing in mind the need for the blood to flow through the muscles. The harder they are contracting, the harder they are to refuel. It is also an interesting parallel with the internal combustion engine where higher power also means more revs. But this factor cannot justify the very large differences in gearing that are common.

I have known riders who would happily ride a gear of 110" in a 25-mile time trial but who for a 4km pursuit put on a 92". The increased speed over the shorter distance would have increased the pedalling rate more than sufficiently. Graeme Obree claimed to have ridden everything from his record breaking 50-mile time trial to his gold medal 4km pursuits on the same 114" gear. Not that Graeme is your average rider!

But all the evidence suggests that our bodies are quite happy with 100rpm and indeed quite tolerant of a bit either side. A case of our bodies being more flexible than our minds.

Muscle contraction/relaxation ratio

The other factor, the ratio between the muscle's contraction and relaxation, has never been paid the same attention as cadence. This is a shame. Had it been better understood, it could have saved many people the effort they quite pointlessly put into making a whole lot of totally useless devices.

These were all intended to improve the humble bicycle but had the exact opposite effect. I have to tell you, they are still out there, and as I now work for the world's largest performance cycle manufacturer, I am like a magnet for them. Please, if you know an 'inventor', give him this book for Christmas!

It has been widely thought by these inventors and many cyclists that pedalling as such, and even the pedal/crank arrangement, is not a very efficient way of extracting power from our bodies. They reason that, as the leg can only push the crank round really effectively for a small part (about 60°) of its rotation, at least 120° of potential power has been lost. So what we really need is a system that allows for 180°, or the whole of the leg's extension, to be used for power production. Maybe a system that allowed you to 'pull up' as well could use up some of the other 180°.

Up to about 100rpm, spinning beats pushing for max power.

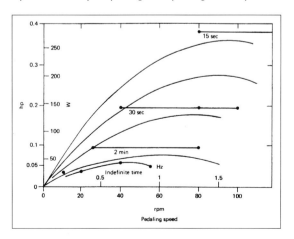

Radical bike design and the legs to match. Graeme Obree broke the world hour record on his home-made special.

All of which is to think just like a mechanical engineer, something that I have to admit I had done for many years. What spoils this very nice theory, quite apart from the ghastly engineering that it always seems to involve, is:

- the need to re-fuel the muscles, and
- the actual optimum ratio of rest to work, which is around 6:1.

That is to say, you will get the most power out of your muscles if they are allowed to rest for six times longer than they are working. So whilst you could push for a whole 180°, as you would not get enough blood flowing through your muscles in the other 180°, you would not push as hard. Equally if you pushed for less than 60° you could push slightly harder, so that although the optimum ratio may be 6:1, it appears to be very much a self-levelling system.

But of course, the 6:1 ratio is almost exactly what you achieve with regular pedals and cranks. They have the added advantages of being cheap, light, easy to fit to most types of cycle, and they are a very good way for our bodies to generate power. They totally avoid shock loads and sudden reversals – rather more benign than running or even walking. The figures for pedal pressures, even for racing, are in the 10-30kg range – just a fraction of body weight – so long as the rider is pedalling smoothly. Contrast this with running, where because of the inertia, loads can exceed body weight.

Use arms and legs to double your power?

Another area of popular misconception that has resulted in inventors giving us a lot more 'improved' cycles takes us back a stage in the power generation process. That is, to where the blood gets its energy. Although Sharp explained it extremely well back in 1896, not enough people read it.

If you have not read Sharp, it seems quite reasonable to make the following deduction:

If a cyclist generates X *amount of power using only his legs, he should be able to produce* X **plus** *power if he uses his arms as well.*

In fact they generally produce X minus power with their awful devices. For short distance anaerobic sprints you do at least have theory on your side, but you have to overcome some serious mechanical problems and some not inconsiderable control ones. Over the longer aerobic distances things get even worse, for not only are the mechanics still against you, but so also is the theory. For here is another parallel between nature and mechanics.

In this case the comparison is with electric motors. You have to imagine that your legs are like an electric motor and that your heart and lungs, pumping and processing the blood, are like a battery. To start using your arms too would be like plugging an extra motor into that battery. This would make sense if the battery were of unlimited capacity. However, as most of us are only too well aware (especially as we wheeze up a particularly steep climb with our heart rate close to maximum), we do *not* have an unlimited capacity.

True, we can develop our heart and lungs with training. But we also develop the muscles in our legs to more than keep pace. And, as in the mechanical world, it is better to run one engine close to its maximum than two at lower rates. For sprint purposes, where the white tissue dominates, this does not apply, and tests on ergonometers *have* shown useful power increases. But in the *real* world it has never come close!

So, to bring all budding inventors down to earth, you can never 'double your power' by any means. A trained cyclist already manages to get something like 95% of all the power that his body can generate into the rear wheel of his bike. This is not to say improvements cannot or will not be made, but they will be of the order of 1%, not 10%.

Hot but not too bothered

I said at the start that our bodies generate power in a similar way, and at comparable efficiency, to internal combustion engines. About 25% of the fuel consumed is converted into useful energy, and both 'engines' dissipate the remaining 75% as heat. Some of this heat is quite useful, even essential. But when flat out most has to be got rid off or the engine would suffer.

Fortunately nature has given us quite a good built-in radiator in the form of skin and sweat. This, coupled with our use of clothing, makes it relatively easy to maintain a comfortable and efficient body temperature. Most of the time, that is, for the amount of excess heat generated by a racing cyclist is considerable. This is not generally a problem because of another problem. The air that prevents him going any faster is, at the same time, doing a useful job conducting away all that excess energy. Try to operate at the same levels on an indoor trainer and you will soon see your power slipping away in a pool of sweat.

The same sort of problems face the riders of fully-faired HPVs (Human Powered Vehicles). The author's own experience suggests that in the more temperate climes only a small airflow is sufficient to maintain power, even for racing over long periods. However, at higher temperatures (25°C or more) and when struggling on big hills, it can be a very big problem. Some work still needs to be done in this area to improve ventilation.

Ergonomics – Sitting Comfortably?

Ergonomics is broadly speaking the science of designing things to fit people. It's a newish idea, as in the past people were expected to adapt to things.

So ergonomics would seem to be just the thing for us cyclists. With it we could have bicycles that fit our awkward little bodies, rather than having to get used to a funny crouch and a saddle that always looks too thin and hard.

There are, though, a couple of problems with this. Firstly, adapting a thing to allow for a perfect interface with people could well compromise the thing's function, which in the case of the bicycle

Comfort is not always obvious. If you are popping down the shops for a six pack then the 'sit up and beg' will work fine, indeed it is hard to beat. On the other hand if you are planning a 100-mile time trial not only would the sit-up-and-beg's aerodynamics be a big disadvantage but even the initial comfort would soon be gone. The problem with sitting upright is that your weight is taken by your backside, which is not good. You also tend to use a wide saddle because of this, which in turn makes pedalling for long periods most uncomfortable. Far better to lean forward take some of your weight on your arms, and as you will be pedalling fairly briskly (it being a race), some of your weight is supported by your legs. Not that a time trial is ever as comfortable as popping down the shops for a beer. You can have it both ways, of course; it's called a recumbent, but that's cheating, isn't it?

could be a poor trade-off. More importantly, what may appear at first sight to be 'ergonomic' and comfortable (such as a large soft saddle) often turns out to be not so good in the long term. The best solution is, of course, to adapt the thing where it makes overall sense, but to be able to adapt ourselves where necessary.

None of which is of much direct help to cyclists. But I hope it will give the reader an idea of why it is so difficult to be scientific about your exact position on a bike. This has not stopped a lot of nonsense being written on the subject of riding position. A lot of it has been written with remarkable conviction, but none of it with any real scientific back-up. Not that I am offering the reader any real science either. But then I am not pretending to.

The one definite scientific fact I can tell you is that nobody can tell you exactly how you should be riding a bike. That is, not unless they have spent a long time measuring you and studying your style. We are all individuals and that includes our comfort needs and pedalling style. And, at the end of the day, each rider has to make the final choice of position for him- or herself.

I said that I had no science to support my arguments. But I can offer some common sense for you to consider. It is obviously difficult to get any dimension exactly right when setting up a bike. *So what is important is to know on which side to err.*

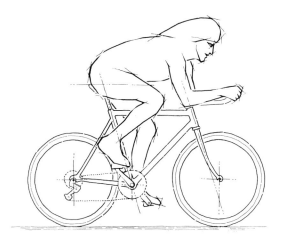

Pedal-to-saddle distance

The most critical dimension for the rider of a safety bike is the pedal-to-saddle distance. There are several formulas for calculating this distance. None of which you need to bother with, as they can only generalise. All you need to do is set the saddle so that your leg is fully stretched with your instep on the pedal. Your normal foot position, with the ball of the foot on the pedal, then gives a slight bend in the knee with the foot in the lowest position. This system (if you take into account shoe plates, sole thickness, etc.) will get you as close to your correct height as can be done by 'formula'.

You will then have to fine-tune this position to suit your style of riding and your exact needs, whether racing or shopping. And for racing at least, this should be done on the basis of the saddle being too high rather than too low. This is definitely a good thing for your body, as it is not a good idea to put a lot of pressure on the knee when it is well bent, as it would be with a low saddle. It is also likely to help you achieve a more aerodynamic tuck for racing. And it is *probably* better for power generation, as the leg can exert more thrust when nearly straight than when well bent.

So, not quite science, but it seems to make sense. The only argument against high saddles is the difficulty of getting a foot down at the traffic lights. For this reason shoppers and even tourers can compromise a little here. They seldom need maximum power output, and a lower saddle may be a little more comfortable for some.

Saddle choice is the most personal thing – and defies all logic in many cases! My only advice is that they should be thinner and harder the further and harder you plan to ride.

Saddles are a matter for personal taste;
choose one to suit your bum.

Crank length

Closely related to saddle height is crank length. And for the same reasons that a high saddle is better than a low one, short cranks are better than long – better for your body and for your position. And, of course, they are lighter!

Traditional Lengths

What the target length should be to start with is rather more difficult. Tradition has us riding 170mm for the average height male. That is, except for mountain bikes, where it is 175mm. Because mountain bikers have longer legs? No, because longer cranks give more power. Do they really? So why don't we fit them to all our bikes? Er, um, because when you are riding off road you pedal slower. Really? And why should you do that when you have 24 gears to choose from? And of course, that is exactly what changing crank length is the same as – changing gear. Whatever crank length suits you for the road will also do fine on the velodrome and the mountain.

I know it sounds wrong and that longer cranks will be easier to push down for any given gear. But power is a combination of *torque*, which is the pressure you can apply, which is greater with *longer* cranks, and *speed*, or how fast you pedal, which is greater with *shorter* cranks.

Ultra-Short Cranks

I am not sure where we got our 170mm dimension from in the first place. Until recently I had no problem with it, other than a preference for 165mm for serious racing. But things are stirring in the world of recumbents. A shorter crank can mean a smaller fairing, and thus lower drag and a higher speed. Pioneers in this area are John and Miles Kingsbury, who re-invented a device that was originally (and of course pointlessly) intended to increase the crank length on the power stroke and reduce it on the return stroke.

The Kingsburys reversed this, so that they got a

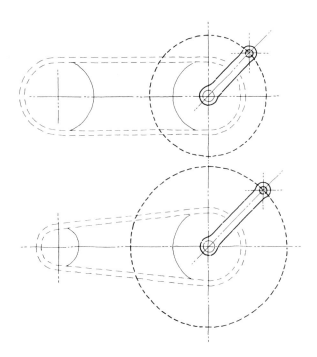

No free lunch. If you lengthen your cranks you will indeed be able to create more force or torque and thus push a bigger gear. But there is a price to pay for this extra torque; your foot will have to travel correspondingly further than with the shorter crank, and this is the bottom line: the total amount of work done is a function of this distance travelled, it does not matter especially whether it was one slow revolution or two fast ones. In practice there is a correct crank length for each leg length, i.e. 170mm for the average person. You can usefully use shorter ones but never never longer.

Engineering or what? Don't let this K-drive from the Kingsburys put you off building an HPV. There are easier ways...

If you don't fancy building a K-drive, Frank Lienhard demonstrates the easy way of doing it: cutting your cranks down so they're just 110mm long.

very short (60mm) crank when on the vertical power stroke and a long (perhaps 220mm) crank in the horizontal mode. The main drawback of this device is its sheer complexity. But then the Kingsburys have this CNC[1] milling machine to play with.

A rather more practical approach was taken by a young German, Frank Lienhard. He turned up at the 1996 HPV Eurochamps with an ultra low-rider with a diminutive fairing made possible by a pair of 110mm cranks. These had simply been cut down from regular cranks. Being just one element of a complex faired machine, it was hard to assess the effectiveness of the cranks as distinct from the aerodynamic advantage. But I met Frank again recently and saw him riding with the 110 cranks on a conventional short-wheelbase recumbent where there was no aerodynamic advantage. He was still able to perform very impressively, even off-road!

As using ultra short cranks would help me achieve a more aerodynamic riding position also on my 'safety', I experimented with 125mm on my work bike to see if I could get used to them. (After all, we all rode on 100mm or so on our 'high wheelers'.) But I gave up on them. I could never get used to them as all my other bikes are 165-170mm. But my son, who has a problem bending his knee following an operation, has been using them for several months now. He is completely happy with them.

---

[1] *CNC = Computer Numeric Controlled.*

Saddle-to-bottom bracket relationship

Traditional advice

Without doubt the least scientifically defined dimension is the fore and aft position of the saddle in relation to the bottom bracket. This is largely determined by the seat angle and frame size, but can be altered by moving the saddle in its clamp, and even using different seat pin designs. The classic advice here that everyone seems to quote is that the knee should be directly over the pedal spindle when the crank is in the forward horizontal position. I have tried to find out where this advice first came from but without success. Whoever it was obviously had never heard of recumbents and knew little of mechanics, as cranks rotate through 360° and are quite happy to be pushed from any direction.

The only difference I can see that moving the saddle fore and aft can have is to alter the angle between the upper leg and lower body. And if this does make a difference, what about the change in this angle that occurs when the rider goes from a low tuck on the drops to sitting up on the tops to climb a hill?

Not that we should be creating any angle at this point, for when using our muscles to propel us in our natural mode we are completely upright. And although this may not be how we were originally designed, we have been out of the trees for a long time.

Fast forward

So, forgetting the nonsense, I would suggest you try riding with your saddle as far forward as is comfortable for you. This will not only mean different angles for different people, but also for different uses. Where performance is the main consideration and aerodynamics matter, you need to get your back as low as possible. This is never *comfortable* but can be made easier by moving the saddle forward. There is then less of an angle between back and legs. This will also have the effect of reducing the amount of weight taken by the saddle and, of course, increasing the weight on the arms.

Seat angles

From my own experience you are unlikely to be happy with a seat angle steeper than 75° if you are using regular drop or cow horn bars. But if you are using elbow rest/tri-bars you can easily go to 78° or more. You may experience, as I do, some niggles with your saddle at these steep angles. This seems to be because, with the nose of the saddle tipped down two or three degrees, you then tend to slide off it. Hopefully it will not be too long before someone designs us a proper 'tri' saddle.

For touring, where comfort tends to take precedence over performance, it is simply a case of getting the balance between the weight on your arms and backside. Shallower angles mean more weight on your rear quarters and vice versa. For city use and short trips shallow angles in the range 68-70° are in order, coupled with a larger saddle. This gives a more relaxed position and makes it easier to put a foot down – a useful idea for kids' bikes, too.

Obviously many of these suggestions are a bit radical and I can imagine that many experienced road racers would say that it is impossible to climb any distance with a 75° seat angle. But that, I would argue, is because they have never tried it – or at least, not for long enough. For your body does get used to generating power in a given way, and the less variation it has to cope with, the less adaptable it becomes. So whereas your typical pro ends up totally inflexible, I am able to ride anything on wheels (slowly!). So please do give it a fair try.

I am inclined to think that the same angles could also be used on mountain bikes, except that handling and traction could be affected. So experiment by all means, but don't blame me if you fall off!

Saddle-to-handlebar dimension

Time trial
Rather less contentious is the saddle to handlebar dimension. This is generally established by holding your elbow against the point of the saddle and estimating the gap between your outstretched fingers and the tops on the regular drop bars. This can be anything from 38mm for the classic British time trial position, to 75mm for the modern 'long road' position. I have tended to go with the fashion here, as my 1981 TT bike is set very short, whereas my current monocoque is a full 75mm. So ride what you are happy with but do give any position a fair trial before changing.

Triathlon
For tri-bar use I can be rather more helpful. The first thing to understand about the elbow rest position is that you still need to get your back as low as possible. A lot of people simply bolt tri-bars to the top of their regular bars. This is often much too high, and any aerodynamic benefits from better arm position can be lost through higher body position. So first make sure that you have enough adjustment to go beyond your normal

The aero time trial tuck is not necessarily as uncomfortable as it looks, though you'll likely need to experiment with it.

position, just to be sure. You should then be able to get the body well down. You then have a choice of positioning the elbow supports low and well back for the traditional 'tri' position, or higher and further forward. This latter position is currently known as 'Superman' and has been used to good effect by Graeme Obree. It is now, of course, banned by most cycle sport authorities!

But you can go a *bit* forward and up, which will help aerodynamically. This is because, for any given body position, the higher the arms are, the lower the frontal area. You also need to decide how to set the spacing on the bars and rests. If, like me, Mr Boardman and a few others, you have a genuinely 'aero' seat post, then you should aim to have your forearms parallel to the ground and each other, with around 100-125mm gap between them. This will minimise frontal area and the seat post will allow the air to flow through cleanly.

If, on the other hand, you have a regular round seat post, and especially if you ride a 'full size' frame with a seat cluster right up in the airflow, then you should adopt the more traditional arm position with hands together, elbows out and up about 15°. This will deflect the maximum amount of air around you.

Handlebar height

General advice
Back to regular bars and setting their height. This is probably the easiest thing to do. Lower is more aerodynamic, higher is more comfortable. You pays your money…

Off-road
For off-road use the same general principles apply, but control has a higher priority. Saddles are often set a few millimetres lower to allow for easier body movement. (A few *inches* lower for downhillers!) Stems should not be too long. Equally, over-long wheelbases are probably not ideal, as you need some weight on the front wheel for control. And avoid high top tubes for obvious 'personal' reasons.

Recumbents
It would have been nice to have included a section on recumbent riding positions, as I have been working with them for so long. But even after all those years of lying on my back I realise that I still don't know of any golden rules. There is still so much experimenting going on and so many riding positions to choose from that you will have to do what everyone else is having to do – experiment for yourself!

3.

Handling – Positively Round the Bend

I have to start this chapter by confessing that I have no worthwhile experience of its subject matter. This means that a lot of the information is essentially theoretical. I hope nonetheless that it is useful.

The reason for this rather detached approach is that I really am a terrible cyclist. Not only am I old and slow, but any bend in the road sharper than the M25[2] has me reaching for the brake levers. This, of course, makes it difficult to speak with any authority on bike handling. Not that a lack of experience ever stopped my journalist friends pontificating endlessly on the subject! So this is my ten cents worth…

Lessons of a misspent (motorised) youth

It is not that I can't corner quickly, it's just that I can't do it on two wheels. Give me a Speedy recumbent trike and I can corner like a Finnish rally driver. This is a technique I learned in my misspent youth, when I worshipped at the altar of the motor car. I was a most able disciple – a fully paid-up member of the four R's club (Raced, Rallied, Rolled and wRecked). I also put them back together, and eventually rebuilt a Healey Sprite for circuit racing. This is not a path I could recommend to budding cycle designers. It did, however, teach me a lot about suspension and handling, some of which is relevant to cycle design.

Mike Burrows 'cornering like a Finnish rally driver' in a faired Windcheetah recumbent.

---

[2] *For the benefit of non-UK readers, M25 is the London orbital motorway.*

The first rule of going round corners fast is to keep the wheels firmly on the ground. Your cornering power is a function of:
- how much rubber you have on the road,
- how soft/grippy it is, and
- how hard it is pressing down onto the road surface, as compared to how hard the car's weight is trying to push it sideways due to the centripetal (cornering) force.

This is why many racing cars have wings that create extra down-force but add very little actual weight that would contribute to the side-force. But even with wings there are still problems to overcome. For what may look like a perfectly smooth road surface is, at the speed a racing car travels, a series of bumps and ripples. Without a very good suspension system the wheels could either leave the ground completely over the more extreme bumps, or at least have the down-force reduced. Either way the result is a tendency not to go round the corner but into the barriers.

An additional complication for car designers is the large and powerful engine that usually only drives two of the wheels. These generally need larger tyres to cope with the additional forces created. And if this was not enough, there is still the individual's style of driving to allow for. So a lot of time is spent at a race meeting setting up the car to suit particular corners, etc, as each driver sees it. Change the driver and the poor mechanics have to go back to square one.

S low and underpowered

All of this is necessary because in motor sport cornering is a very important part of race performance, especially with races being won or lost by seconds. Cycling is not like this. Firstly, bicycles are really really *slow* by comparison. Therefore what looks like smooth asphalt is smooth asphalt and our annular pneumatic suspension[3] is more than up to the task of keeping the rubber on the road in anything but the roughest of town centre 'crits'[4] or Belgian cobbles. Also our 'engines' may well be clean, green and mostly silent, but they are totally useless in the power and output stakes. They would have difficulty running a decent sound system, let alone burning rubber. So there is no need for extra large tyres on the back – you are not going to spin off at the hairpin.

---

[3] *Tyres – see later chapter.*

[4] *Criteriums = city centre races.*

Get a grip

What does matter mostly is how much rubber we have in contact with the road and how grippy it is. And it is this grippiness that is probably most important. Yet it is the thing that we know least about, as none of the tyre manufacturers I am aware of offer a choice of grades as is the norm for car and motorcycle racing. Some have started to offer a range of colour variations, the hue of the side stripe indicating its suitability for various conditions. I shall leave it to the reader to decide whether these pretty colours are the result of science or marketing.

You can of course use common sense and simply feel the tyres. Or, if blessed with more courage than me, carry out your own tests. Don't forget the leathers!

My own favourite road tyre is the Michelin Hi-lite. Never taken to the limit as such, except on the back wheel of my Speedy, but they do wear reasonably fast, which suggests that they are gripping well. They also come in 'slick' format, tread being the last thing a road tyre needs. For as I said, what you need is rubber in contact with the road, and a tread means gaps in the rubber, and gaps mean no grip at all. Even the edges of the tread do very little. What does provide the grip is the edges of the road – all the small stones, etc. that make up the texture of its surface. They dig into the rubber of your tyres and stop you sliding about.

You never see racing cars or motorcycles with tread any more. Even in the wet they don't use tread – just a softer grade of rubber with big grooves to let out the water. The more modest cross-sections of cycle tyres do not require such drastic treatment to keep them in contact with the road surface, even when it's wet. Getting hold of softer compounds is more of a problem, especially as most professionals and top amateurs prefer the glued-on tubular tyres, which seem to be on the hard side. The tradition of keeping them for six months before use can only make matters worse.

Big flat bottom

The most significant factor in putting rubber on the road is, of course, the size of the flat bit on the bottom of the tyre. This is a function not of the size of the tyre but of how much pressure you have put in it. It is obvious when you think about it, as my friend Mr Kingsbury pointed out to me not all that long ago. If you have a total bike and rider weight of 200 pounds, an even weight distribution and you inflate the tyres to 100psi, then each tyre will have one square inch of rubber in contact with the road. If it is a narrow section tyre there is not much grip to be gained by lowering tyre pressures if you are to avoid rim damage due to bottoming out. However, we run larger cross-section tyres at lower pressures and thus do get more grip.

Typical city tyre: loads of rubber, loads of grip.

Myth and legend

Those, then, are the factors governing the amount of grip your tyres produce and the consequent cornering power of your bicycle. This elegant and relatively simple picture is rather spoilt, however, by the rider. For, like the racing car driver who has an ideal set-up, our rider has his likes and dislikes. But whereas the car driver's requirements are at least mostly based on science and common sense, the cyclist's wants are more to do with myth and legend.

Not that I can out-ride any of them, but I have done quite a bit of thinking and experimenting, and certainly have experience of more varieties of cycle than most. In particular, I have built many recumbents, both two- and three-wheeled. In so doing I have had to do some *real* design, not just copying. One machine in particular, the Ratcatcher, was most illuminating. I had built an earlier recumbent and had not been very happy with its handling. So I set out to find out for myself what really made a difference.

A different angle

While I was pondering these problems a friend loaned me a copy of *Motorcycle Chassis Design* by Foale & Willoughby, now sadly out of print. In this the authors pondered, just as I had, 'why do we have a given head angle?' and 'would alternatives work?'. They were not building recumbents but chose a BMW 650 to adapt. Motorcycles normally have a head angle of 28° (motorcycles are measured from the vertical) so they modified their bike to 15° and also to 0° – dead upright, and with the forks raked back 2" to maintain the correct trail. To their surprise the 15° head angle was an improvement, but best of all was the 0° set-up. The bike was better at low speed, with no bump steering on rough surfaces, the suspension worked better, and it could be ridden at 100mph hands off! The only negative effect was brake judder due to the forks not being designed for these loads.

I set up the same experiment myself using my Ratcatcher long-wheelbase recumbent. I arranged for a steering head that could be set at any angle from 45° (very shallow) to 80° in reverse. I combined this with a pair of forks that could be set at any rake, forward or reverse.

My intention with the Ratcatcher design had been to produce an easy-riding touring machine. I had assumed that the more laid back angles would give the desired characteristics. Sure enough, initial impressions confirmed this. But on further riding and analysis I realised that I was having to make a lot of small corrections just to go in a straight line.

For those not familiar with recumbents I should explain here that the ratio of steering to leaning is higher than on a safety bike: there are not many recumbents that can be ridden hands-off. So I rode and rode with every angle/rake combination you can imagine. And I got everyone I could to act as guinea pigs as well. The outcome of all this was, not surprisingly, the same as Foale & Willoughby's discovery – an upright head and forks raked back about 22mm on a 17" diameter wheel. This gave very positive and accurate steering with no sense of twitchiness.

The reasons for all this are well explained by Foale & Willoughby. Put simply, a wheel with vertical steerer will, when turned, have a 100% effect on the direction of your bike, but no effect on its height. In contrast, a horizontal steerer will have no effect on the direction but 100% on height. Therefore, any angle other than vertical will compromise the steering to some degree. It does get more complex but you will have to buy their book if you can find a copy.

Foale & Willoughby concluded that the reason we use the angles we do is less to do with steering dynamics than ergonomics. We need handlebars in a certain place and a given wheel position, so the angles used both by cycles and motorcycles are a compromise. But they are probably a *good* compromise and there is no real reason to challenge them, except in the case of recumbents. It is nonetheless important to understand this, otherwise you are likely to end up being obsessed with angles and rakes, the fine detail of which is totally unimportant.

While I was carrying out the tests on Ratcatcher, I was having to change the angle by at least 10° to notice any real difference in the handling. So I would argue that for the range of angles that we tend to use in cycle sport (say 69° to 76°), and assuming the correct fork rake, differences will be mostly in the rider's head. The more experienced rider will be more likely to notice small changes. But even if he does, there is no reason for these

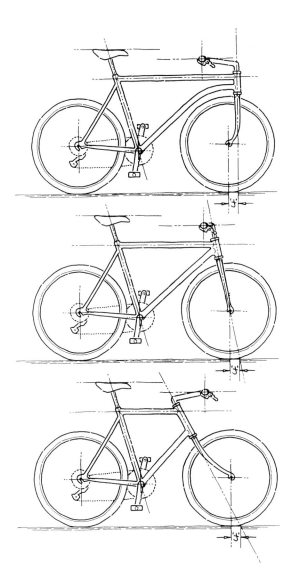

Trail is the main factor in determining the feel on handling of a bike, and there are good and bad ways of getting it right. Option A, although very silly for a 'safety', is actually a very good option for recumbents.

changes to affect the bike as such. It is a matter of feedback, and as long as the rider understands this, he can ride road race, velodrome or probably even off-road on the same head angle.

Weight distribution

The other factor I hear mentioned is weight distribution, which in reality for road racing is very much a self-compensating thing. This is because if you have more weight over a given wheel, you will not only have a greater force acting at an angle to the direction of travel, but also more grip to compensate. And for safety bikes we are talking about very small changes. If you want to see this effect in its most extreme form, you will have to visit an HPV race. There you will see:

- long wheelbase designs with 75% of the weight on the rear wheel,
- short wheelbase with 65% on the front wheel, and
- the occasional medium wheelbase with something like a 50/50 split.

Any difference in cornering speed you will quickly realise is due to rider rather than machine.

Conclusions

So to sum up, you want good tyres with no tread. Size, as they say, is not so important but pressure could be. Wheelbase is not important and neither are angles or fork rake. What does matter is *you*. There are no good bikes, just good riders. And I should know, because I'm crap on all bikes. So if you can ride a bike, and as long as you don't let your mind get stuck in a rut, your wheels never will.

All of that lot was for roadies, of course. There are other areas of bike handling where I have even less experience.

Velodromes

There are no corners on a banked track, so why worry about angles at all? I reckon you could ride anything on a velodrome. Just avoid heavy monoblades on whippy frames, as they shimmy if you ride hands off. And the one thing I know about velodromes is that you have to be able to ride hands off to look cool.

Off-road

Much the same as road, except you mostly don't want slicks, and size *does* matter. And even more crap is talked by road testers about angles and how 71° is a bit laid back and good for downhill sections, whereas 72° is a more technical angle for climbing and single-track.

So how come we use the same frames with anything from rigid forks to four-inch travel triple clamps that change the angle by 2 or 3° and don't have a problem? Because it's mostly bo**ocks, that's why! There are angles and rakes that are better than others but the difference is very small, especially when you are riding in mud or loose rock. So stop making excuses about your frame, get out there and ride it. Unless it's getting close to end of the season, and you are looking for an excuse to buy a new toy. If so, you will find just what you want in the new Giant catalogue.

Trikes

Funny things trikes. Only taken seriously by the Brits and then not too seriously: most cyclists either love or hate them. I started my cycling career on a borrowed Higgins tandem trike, so I obviously quite like them. However, all my current trikes are the recumbent variety, hence this extra bit of thinking for you.

Regular trikes may well look a lot like regular bikes, sharing as they do wheels, fork and most of the diamond frame. But this is very misleading. They are, in fact, a breed apart. For one thing they never do (at least if you are lucky) is lean on corners. Therefore to have a steering geometry related to the bicycle would seem to be nonsense.

Far more sense to relate to the motor car, as I did with the Speedy. This has kingpins angled back at the top 1° (head angle) with the axles on the kingpin centre lines (straight forks). Anyone interested in a 'barrow' that does not follow every rut and bump in the road, and does not need to be constantly dragged back up the camber, could do worse than put this theory into practice.

4. Materials and Processes

*A*nyone who regularly reads any of the popular cycling magazines will already have some idea of the properties of the various materials used for frame construction. Unfortunately it will be the wrong idea.

This is because the various reviews and road tests contained in these journals are inevitably written by journalists. Such articles are intended to be interesting and entertaining, which is, I suppose, as it should be. Sadly they are seldom written by cycle designers. And they are almost never written by engineers, most of whom would know better than to perpetuate (even in the name of entertainment) the many harmful myths that surround cycle frame design and construction.

The various materials used in frame construction do, of course, have different qualities. It is important to have a good understanding of them when designing cycles. But the actual characteristics are usually only measurable with sensitive instruments – and almost never with the posterior of a homo sapiens! Even more importantly, any differences that may exist in the final complete frame will be more to do with its design (and specifically the tube size) than the material chosen for its construction.

So then, here are some of the various materials and their properties that you need not bother looking for in your own bike!

Steel

Almost universal, some 95% of the world's bicycles are made from steel tube. And for very good reasons. It is cheap, easy to process, durable if well painted, and the hardest, stiffest, strongest, structural material available. It varies in tensile strength from 375 to 1800MPa[5]. This is a measure of its breaking point if simply 'pulled' like a piece of string. Its modulus of elasticity though, which is the measure of its stiffness or resistance to bending, is the same for all grades – about 201,000MPa. Anyone claiming that their super-high-strength tube-set will give a stiffer frame is lying.

---

[5] *MPa stands for MegaPascal. 15 MPa approximates to 1 ton force per square inch.*

Mild steel

Most steel frames are built for people who have no interest in their modulus of elasticity. Such frames are made from thick-wall tubing that has been rolled from mild steel strip and electrically seam-welded. This produces a strong, stiff, cheap, durable and very heavy frame. Most of these frames are produced and used in China and India – or given to Western children for Christmas.

Low carbon (Hi-ten)

Not much more expensive, but a lot better, is low carbon steel. This is often referred to as Hi-ten by the cycle industry. It comes either seam-welded or cold-drawn and is the basis for most of the West's 'bread and butter' bikes. Where weight is not a major consideration it is a good choice. The money saved on the frame can be put into better components, giving you a more reliable bike.

Chrome moly (4130)

However, the moment you start talking about performance, weight will start to matter. This means a 'low alloy' steel, usually referred to as chrome moly, or 4130, which is its aviation industry specification. Indeed, it was developed for aircraft structures, but that was in the days of biplanes. It is still very useful for cycles, and most of the world's 'affordable' performance bikes were built from it.

High quality tubing – from 531 to Airmet

Chrome moly is not so commonly used by the many small frame-builders around the world. They spend a lot of time building a frame. The material cost is less important than their labour, which is much the same for regular 'chromo' as it is for one of the fancy-numbered tubes.

At one time the only number a cyclist needed to know was 531. It was the universal tube for racing cyclists worldwide. Not so today. Reynolds, the British producers of 531, have another six or so. Columbus in Italy have a similar range. And there are French, Japanese and American products to choose from. They all tend to offer

Tubes are butted because you weld at the thick bit at the ends.

TYPE	COMPOSITION %								MECHANICAL PROPERTIES		APPLICATIONS etc.
	C	Si	Mn	Cr	Ni	Mo	W	V	T N/mm² or MPa	Elong. %	
Low alloy structural steel	0.3	0.3	0.75	–	3	–	–	–	800	26	Crankshafts, high tensile shafts etc.
Nickel-chromium-molybdenum steel	0.35	0.3	0.7	0.8	2.8	0.7	–	–	1000	16	Air hardening steel. Used at high temperatures.
High tensile steel	0.4	–	–	1.2	1.5	0.3	–	–	1800	14	Used where high strength is needed.
Spring steel	0.5	1.6	1.3	–	–	–	–	–	1500		

W = tungsten V = vanadium

Different grades of steel have different properties, although the modulus of elasticity (which measures stiffness) is the same for all grades.

similar types of alloy giving progressively higher tensile strengths. This allows the wall thickness to be reduced, keeping up the frame's strength but reducing weight and stiffness! Top of the pile at the time of writing is 'Airmet' from the USA. At around 1500MPa tensile strength it needs very special machining and welding techniques.

Butted tubing

Virtually all of these specialised tube-sets and a lot of the chromo comes in 'double-butted' form. That is, the last 50mm or so at the end or ends of each tube is some 40% thicker than the centre section. This is not because the stress is higher at the ends but to allow for a possible loss of strength caused by the joining process.

Typical tube diameters

Steel tubing is almost always circular in cross section with little variation in size. Traditionally the down and seat tubes are 1⅛" outside diameter, and the top tube 1". And no, I have not forgotten that we went metric, because cycle tubing hasn't. On mountain bikes there is a degree of oversizing, usually up to 1¼" and 1⅛" respectively. This trend has even appeared in some fancy road tubing, along with a degree of ovalising, notably Columbus 'Max'.

Lugs and brazing

The joining of steel tubes is something that the industry understands very well. After all, it has been doing it for 120 years.

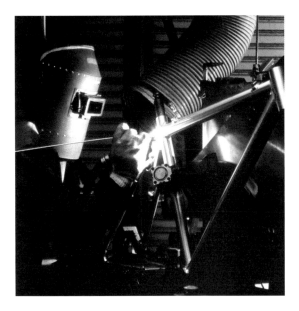

TIG welding – still craftsmanship, even if it's done on a production line.

Commonest process still is using lugs and brazing, with a regular brass filler. This can either be automated using various techniques or done exquisitely by hand. In the latter case a silver alloy may be used in place of the brass to reduce the heat input to the tube, so retaining more of its strength.

MIG welding

Faster, cheaper and increasingly used for volume production is MIG welding. Metal Inert Gas welding uses a filler rod of similar material to the frame. This rod is fed through the nozzle of the welding torch surrounded by a shield of inert argon gas. The arc that flows from torch to frame fuses the tubes and filler into one.

TIG welding

MIG is an ideal process for automation by robots and produces quite neat results. But not on the thinner wall tubing. For that you need TIG – Tungsten Inert Gas – where a fixed tungsten tip replaces the consumable steel filler rod. The resulting arc is very controllable and almost like an intense gas flame. Clever people can weld kitchen foil with TIG!

A filler is still usually used to fill the joint, in the same way as brazing. This is the preferred process for high-volume production of chromo frames. It even has devotees in the world of handbuilding. Most of the modern high strength alloys can be 'tigged' successfully. But long term fatigue may not be as good as traditional lugged construction.

Fillet brazing

Slowest but nicest is fillet brazing, using an oxyacetylene torch and a bronze-based filler rod. A large fillet is built up, rather than a lug, so providing some stress distribution at the joint. The fillet then has to be filed, sanded and polished smooth – a very slow process. But the appearance is oh so elegant. Look at a Chas Roberts some time…

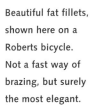

Beautiful fat fillets, shown here on a Roberts bicycle. Not a fast way of brazing, but surely the most elegant.

Aluminium alloy

Although only a small percentage in global terms, aluminium frames have come to dominate the performance scene. Aluminium has a lower tensile strength than steel, typically 225-750MPa, and a considerably lower modulus of about 54,000MPa. It is much more prone to fatigue if badly used. But it is a third the weight of steel. This is why it is what most aeroplanes are made of today, although not in tubular form.

All structural aluminium is alloyed with small percentages of other elements, in the same way as the high grade steels. It is designated into groups that share a similar main additive. They run from the 1000 series which is pure, to 8000 which has lithium as the main addition. Higher numbers are not necessarily better, and grades of one series can be stronger or weaker than those of other series.

6061 and variants

The commonest grade for the cycle industry has been 6061. This is magnesium and silicon-based, is quite strong at 420MPa, can be cold formed, has good corrosion resistance, and welds well. However, it requires specialised heat treatment afterwards, or it will crack at the joint. This involves heating the whole frame to 450°C for one hour, quenching in cold water to fully anneal the welds, then reheating to 140°C for 30 minutes to achieve the famous T6 status. For that is what T6 is – a condition, not an alloy as such. Various aluminium heat-treatable grades can be T6.

My own employers, Giant, use a very interesting variation, officially designated 6013 but known as Cu92. It has a high percentage of copper, normally used only in the 2000 series. This gives a very useful increase in strength but does not affect its other qualities – other than price! The 2000 series themselves can be high-strength but are non-weldable, and prone to corrosion if unprotected.

7005 and 7020

The other big favourite, especially with the mass market producers, is 7005 or 7020. This is a stronger material than 6061. It shares its formability and generally good corrosion resistance, but does not have such good welding characteristics. It does not lose as much strength as 6061. But even if heat-treated, it can never return to full strength. What happens is called *stress corrosion*. This is cracking caused by the constant flexing that any frame experiences, coupled with galvanic corrosion between the 'grain' boundaries of the weld. The galvanic corrosion is caused by impurities in the form of

Material Designation and Temper	Density g/cm³	Melting Range C	Tensile Strength MPa	Elongation Percent on 50mm	Shear Strength MPa	Fatigue Strength 50x10⁶Hz MPa	Hardness Brinell	Modulus of Elasticity GPa
1050A-O	2.71	635	75	32	50	20	21	69
2014A-T6	2.80	530-610	470	7	293	170	145	74
5083-H22	2.67	580-645	337	8	155	124	95	71
6061 -T6	2.70	570-660	305	9	205	95	90-100	69
6063-T6	2.70	580-660	210	–	155	85	75	69
6082 -T6	2.70	570-660	310	8	218	124	90-100	69
7075 -T6	2.80	475-630	565	–	330	–	150	72

MPa *equals* N/mm²

Typical physical and mechanical properties of some commercially stocked sheet alloys.

dissolved metal salts present in everyday water.

The best way to make use of the high strength of 7005 or 7020 is to produce an extreme form of butting. This reduces the weight but leaves enough metal for a safe weld. This, I suspect, is the reason for the unusual square tubes with machined recesses used on the stylish Pace Research bikes.

5082

The third of the main structural series alloys is 5082, which is very popular with French manufacturers. It has high strength (around 375MPa tensile), good formability, and the best corrosion resistance and weld quality of all alloys. Hence its main use is boat building.

Its drawback is that, unlike the other alloys, it is not a heat-treatable grade. So, whilst the heat of the welding process will lower its temper, it cannot be reversed. The only way it can be hardened is by 'cold working', which is what happens as it is drawn through the final die in the tube-forming process. As with the 7005 grade, extreme butting would be a good idea.

2014 and 7075

If you decide to bond your frame together rather than weld it, you have a couple more grades of aluminium to choose from. 2014, which is one of the commonest aviation grades (about 470MPa tensile), has low corrosion resistance and low formability but machines

well. 7075, which is the strongest grade of aluminium commercially available at around 565MPa tensile, has similar qualities.

8000

The one we are waiting for, but which is a bit slow in arriving, is the 8000 series. The main alloying element is lithium, which is used in the unusually high percentage of 25%. It is of such low density that not only is this the strongest alloy yet, but also the lightest. It even has a higher modulus of stiffness. But it is barely in commercial production and not in tubular form at all.

Scandium

Light as feather, stronger than steel, hard as diamond and faster than a speeding bullet. Well, maybe it wasn't that bad, but the hype that surrounded the launch of scandium alloys made you wonder. After all, if you have a good product the truth is usually sufficient – as it is for scandium.

It was discovered in 1879 by Mr Nilson and named after his native Scandinavia. It is a soft, white metal slightly heavier than pure aluminium, and there is about the same amount of it about as tin or mercury, but spread rather thinner. Until recently the only source was reprocessing old mine workings of other odd metals, and its only real use was for research.

But then in the 1960s the Russians discovered a large amount of it, probably in the form of a thortveitite mineral. Before long they discovered how good it was to mix it into aluminium. As I explained earlier, aluminium alloys are defined by their prime alloying element, usually 1.5 and 6% by weight. But all alloys contain several other elements – titanium, chrome, iron – in very small amounts (0.3%) as a way of refining the grain structure and reducing crystallisation.

It turns out that scandium does this better than anything else, improving strength, weldability, etc. and with no down-side, except that there was none around! Until now, with the end of the Soviet Union, and with US money and expertise to exploit the new wonder product. The tubing currently on offer seems to be a 5000 series alloy, which would normally lose 30% of its strength when welded. Now it loses maybe only 10%, and it is stronger as well. I would guess a lot more is to come as well – just watch out for the hype.

Joining techniques

The bulk of aluminium frames are TIG welded in a very similar way to regular chromo, but with a much larger fillet at the joint. This can

Nothing wrong with glue, so long as you use it properly.

either look smooth and elegant as on Cannondale and Klein frames, or a bit 'functional' as on most others. This is not because of a difference in the welding but down to an individual with a file and emery cloth spending a lot of time on each frame.

Adhesive bonding is still very popular, with Bador in France and Trek in the USA being the largest users.

Aluminium can be brazed but I am unaware of anyone currently producing frames in this way. However, at Giant we add some of the more delicate fittings this way.

Titanium

Nice stuff, titanium. Half the weight of steel with similar tensile strength and a modulus of about 120,000MPa. In theory it is available in a bewildering variety of grades, and in three 'phase' variations. However, I won't go into any great detail about all of this, as in reality you can only get your hands on three types.

Apart from the price, restricted choice is the biggest problem with titanium. It is only available in a limited number of sizes and there is a lack of butted tubing. This is because the tubing that is produced is for aerospace use – not to hold the wings on, but for hydraulic pipes.

Grade 2
Grade 2 titanium, which is 'commercially pure', has a similar strength to mild steel. It welds wonderfully well with no loss of strength, provided you have the right equipment and know what you are doing. And if you ever want to boil nitric acid, this stuff would make a good kettle, as it never corrodes.

3AL 2.5V and 6AL 4V
Better for bike building, though, is '3AL 2.5V' – which means 3% aluminium and 2.5% vanadium as the main alloying elements. This has a tensile strength similar to 4130 chromo steel but still with half the weight, and good welding and fatigue characteristics. Machined drop-outs are often '6AL 4V'. This is even stronger but does not appear to be available as tubing.

Joining techniques
TIG welding is the universal joining process. (Raleigh used titanium in some of their bonded frames – don't ask why!) However, I believe that the British-made Speedwell frames of the 1970s used a fusion process of some sort.

Composites

Black, shiny and expensive is how we normally think of composites. This is because carbon fibre is the commonest of the high-strength fibres used by the cycle industry. It is also very strong and very light. Just how strong, or even how light, is difficult to say. For unlike metals where the 'qualities' of the tube are much the same as for the raw material, the final qualities of a composite are only achieved after all the processing is complete.

Carbon fibre

In the case of carbon fibre, the raw material is in the form of a 'tow' or bundle of very fine fibres – six or twelve thousand to the 5mm wide tow. As with the metals, it comes in grades. Most popular is T300, which is used for the bulk of sports goods worldwide.

Measured in this form as a pure material it has the same tensile strength as the strongest of steels and the same modulus. It never fails due to fatigue and is two thirds the weight of even aluminium. There are grades of carbon with at least three times as much strength as T300, others with three times the stiffness, and some with a combination of qualities. But raw carbon tow is not much good on its own. It needs processing.

Aligning the fibres

Strongest and most specialised is 'UD' – *unidirectional*. Here many tows are laid out side by side. All the fibres run in one direction in a thin sheet, maybe a metre wide.

The fibres are held in place either by a cotton trace cross-stitch or adhesive film strips. Alternatively it can be woven into a conventional-looking fabric using a variety of weaves, all of which will have an effect on the final product. It is even possible to knit the fibres into a three-dimensional 'pre-form', like a sock or glove.

Binding the fibres

In this dry form the carbon still has no useful strength. For this to be achieved the fibres need to

A carbon fibre Giant MCR, designed by Mike. 'One of mine, so what can I say?'

be bound together. The commonest way of doing this is to 'wet out' the fibres with an epoxy resin. This can be done to the cloth or UD fabric and supplied to the manufacturer in 'pre-preged' (i.e. pre-impregnated) form. The resin is cured by heating, typically to 140°C. The 'pre-preg' is stored in a refrigerator until needed.

Alternatively a two-part resin can be mixed and applied to the dry fibres as they are cut and fitted into the mould.

You can then leave the resin and fibre to harden in the mould as it is. More likely you would apply some form of pressure to compact the fibres into the mould and squeeze out excess resin. This is the usual method and is essential for good results.

The commonest way of applying pressure is 'vacuum bagging'. Here the mould is put inside a plastic bag and the air sucked out so that the natural air pressure presses evenly on the whole of the surface being moulded. This can be taken a stage further by using an autoclave. This is like a giant pressure cooker, in which the pressure can be raised to 700kPa[6] and heat applied to improve resin flow and curing time.

For even more control and pressure there is 'matched tooling'. Here a male and female mould are used, usually in steel and with a built-in heating system. It's a bit like pressing panels in the car industry. However, in this case the pressure is held on for 45 minutes or so while the resin hardens. Giant use this process on some models.

Giant use a similar process for the tubes. The layers of carbon and aramid pre-preg are rolled around a plastic core. They are then placed in a heated steel split mould. The heat causes not only the resin to harden eventually; it also makes the plastic core expand, compressing the fibres against the inside of the mould.

So, after all that, you can see that there can be a lot of variation in the final qualities. And that is just using carbon fibres.

Aramid and glass

You can also use aramid, best known by the names Kevlar™ and Spectra™. This is not as strong as carbon but a lot tougher. There is also glass. This may seem low tech but it has a lot of useful qualities, especially in combination with other fibres.

Thermoplastic composites

Then there are the thermoplastic composites. These are usually made by 'co-mingling' or weaving together various fibres, typically carbon and nylon. These are then packed into a steel mould, pressurised and heated to around 360°C. The nylon fibres then melt and become the resin. The resulting material has similar strength and weight to epoxy-based materials, but much better resistance to surface damage. Its biggest drawback currently is the near impossibility of sticking bits of it together or anything to it.

Gluing or 'bonding' (which is the 'politically correct' term) is the only way of joining composites. All the tubular designs use lugs – either cast aluminium alloy or moulded carbon. Examples include Trek's OCLV and Giant's Monex, often (but wrongly) referred to as monocoque construction.

True monocoques, or at least moulded structures of some sort, are the best way to use composites. They bring even more possibilities. These include RTM – Resin Transfer Moulding. This is a bit like the matched tooling process, except that the fibres are put into the mould dry, and the resin is then injected under pressure into the closed mould.

Then there is DMC – Dough Moulding Compound. DMCs are a bit like pizza dough. They can be dropped into a mould and processed in two minutes. Most of this, though, is in the future – but not *that* distant.

So you can see how difficult it is to be accurate about the strength of composites generally. I can tell you that in practice a good carbon epoxy tube

---

[6] *KPa stands for kiloPascal. 100 kPa approximates to 1 atmosphere (atm) or 14.5 pounds per square inch (psi).*

should be stronger and lighter than the best alloy tube. But even that is not saying enough.

The strength of composites is often misunderstood. Carbon/epoxy is a relatively brittle material. When it fails it always does so in a spectacular way – there are no *slightly* bent carbon frames. But the material has such an enormous weight advantage that most manufacturers considerably overbuild their frames. So for any given load (even shock load or impact) the carbon frame should take more abuse than either alloy or steel.

Composite's biggest advantage for cycle construction is its resistance to fatigue damage – that is, the constant small shocks and

(Below) Properties of some thermoplastics.

(Bottom) Properties of some thermosetting plastics.

Name	Density g/cm^3	Tensile Strength N/mm^2 or MPa	Percentage Elongation at break	E N/mm^2 or MPa	Brineli Hardness	Machineability
PVC (rigid)	1.33	48	200	3.4	20	Excellent
Polystyrene	1.30	48	3	3.4	25	Fair
PTFE	2.10	13	100	0.3	–	Excellent
Polypropylene	1.20	27	200–700	1.3	10	Excellent
Nylon	1.16	60	90	2.4	10	Excellent
Cellulose nitrate	1.35	48	40	1.4	10	Excellent
Cellulose acetate	1.30	40	10–60	1.4	12	Excellent
Polythene (high density)	1.45	20–30	20–100	0.7	2	Excellent

Name	Density g/cm^3	Tensile Strength N/mm^2 or MPa	Percentage Elongation at break	E N/mm^2 or MPa	Hardness	Machineability
Epoxy resin (glass filled)	1.6–2.0	68–200	4	20	38	Good
Melamine formaldehyde (fabric filled)	1.8–2.0	60–90	–	7	38	Fair
Urea formaldehyde (cellulose filled)	1.5	38–90	1	7–10	51	Fair
Phenol formaldehyde (mica filled)	1.6–1.9	38–50	0.5	17–35	36	Good
Acetals (high density)	1.6	58–75	2–7	7	27	Good

bumps that are part of a bicycle's daily life. The JIS (Japanese Industry Standard) test for frame life requires a loaded frame to survive half a million 'shakes' equivalent to a 50mm bump in the road before it can be sold. Many frames do not pass this test, but Giant's own Cadex carbon frames survive two million cycles.

Also important, but even more misunderstood, is the ease with which composites can be repaired. To date politics have made it difficult for large companies like Giant to offer a repair service. But increasingly pressure for recycling, or more usefully for re-use, will make it inevitable. After all, it's very easy. The glue is epoxy, which is what the frame is held together with already!

A Kirk Precision, made from magnesium alloy. Not a good idea – not then and not now.

Magnesium

The lightest readily available structural metal. It is not quite as strong as aluminium alloy but quite nice stuff. It is used in aerospace for 'chunky' bits.

The French company Time have used magnesium for pedal bodies, and several manufacturers have adopted it for the lower legs of telescopic suspension forks. It could usefully be adopted for a lot more 'trick' bits. However, it has poor resistance to corrosion, so good surface treatment is essential. Its fatigue resistance is also poor. Hence long bending sections must be avoided.

Somebody said you could cast a whole bike frame in it. But nobody would be that daft, would they?

Metal Matrix Composites

MMC can mean a lot of things. For our purposes it currently means an aluminium alloy to which has been added a small amount (10-15%) of very fine, hard particles. These are usually either aluminium oxide or silicon carbide. The resulting mix is a sort of metallic concrete.

This improves the strength and the fatigue resistance, and even lowers the weight slightly. Weld quality is not quite so good but for some grades is acceptable.

MMC is extremely difficult to machine or even draw into tube.

The Swedish Itera, produced in 1981. Also not a good idea.

Being a sort of metallic concrete, it would be. This is a shame, because using it just for frames is a waste. What we should be producing is sprockets, chainrings, rims, disc brakes, etc. – all of which take advantage of its abrasion resistance.

For the future, Magnesium Electron are developing a magnesium version of MMC. In the 'dollars don't matter' world of defence-related aerospace they are experimenting with titanium and diamond whiskers! And that's just for the ashtrays.

Plastics

Not the composites group but injection-moulded thermoplastics, such as Nylon.

Injection moulding was used to produce a bicycle – the Itera. It was produced around 1981 in, of all places, Sweden. Which is very strange, as Sweden has a well-deserved reputation for stylish and practical good design. The Itera, on the other hand, was conceptual garbage. It was a bicycle so bad that you will have to hunt one down and ride it for yourself to fully understand what 'design' can do for an otherwise excellent product.

The Itera's failure should not stop people thinking about plastics, though. Given a good design concept and some FEA (Finite Element Analysis) work on the computer, it just might be possible to produce something that cyclists, as opposed to design centres, would appreciate.

A good example of progress in plastic is the rear mech. Simplex made a 'plastic' one in the '70s which was horrid! SRAM now do a whole range that are brilliant.

Beryllium

Lighter than magnesium, as strong and as stiff as steel. Sounds too good to be true?

Well, it is true in theory. The problem is that beryllium is a bit tricky to work with. Objects are normally produced from it by a sintering process. That is, the material in a fine powder form is placed in a mould where it is compressed at high temperature to produce the finished component.

Beryllium is almost impossible to machine, weld or bend. It is also highly toxic. I hear that the F1 crowd are now using Beryllium but have realised the dangers and may ban its use. Not a bad idea, as a cyclist with an electric sander could poison his whole family with his weight-saving.

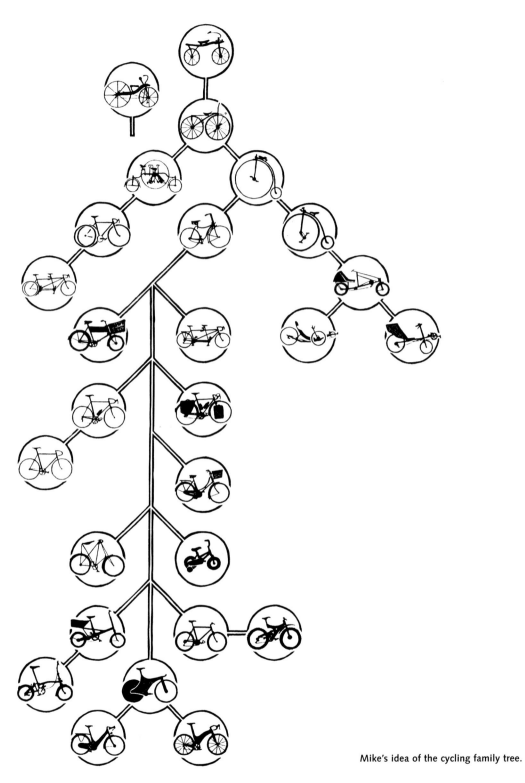

Mike's idea of the cycling family tree.

Materials and Form

Evolution of the Bicycle Frame

Timber beginnings

Way back at the beginning, when von Drais started things off, there was only one practical material for vehicle construction: wood. Today we tend to think of the wooden bicycle as a bit of a joke. This is rather unfair, for although wooden-framed cycles have been rare since the 1860s, there have been some interesting designs.

That these were all failures must have been particularly frustrating for their designers, for wood does possess some very useful properties. It is very light and easily worked, or at least some of it is. It comes in rather more varieties than any metal. And although not as strong as the weakest of metals, its strength-to-weight ratio is better even than aluminium.

A wooden bike: it can be done, but it probably shouldn't.

Wood does have a couple of big drawbacks, though. It has, as we are all aware, a 'grain'. This gives it considerable strength in one direction but very little in any other. Laminating thin sheets in alternating directions to produce plywood gives good strength in *two* directions, but still very little in the third.

Technically wood's strength is known as 'anisotropic'. This contrasts with metals, which have little or no grain and have 'isotropic' strength. Composites, like wood, are also anisotropic. However, as they can be created to give strength just where it is needed, it is less of a problem.

A topiary frame would be just a bit tricky!

Another problem for wood is that, being low density, we use it in solid form. It is not always easy to connect all of the strength of any two or three elements of a timber frame. Gluing tends to bond only the outer layers, and traditional joints tend to weaken the structure. And even if you can solve these problems, there are still areas like the bottom bracket where the stresses are far too high for any wood.

Wood is very nice stuff but on balance is best kept for boats, tables and the like.

Wrought iron: even better for gates than for bikes.

Wrought iron

The fact that wood was less than ideal for making bicycles meant that it had little influence on their form. The next material that the bike builders got their hands on was slightly more influential. This was wrought iron. However, this was as much a step back as forward, because wrought iron was still only available in solid form and was a lot heavier than wood.

Steel tubing

But it was not long before an altogether more appropriate material became available. This was, after all, the height of the industrial revolution. What was about to revolutionise the bicycle was steel. And even more importantly, the technology to turn it into tubing.

The diamond frame

At this point things really started moving. By 1885 there was the first successful rear-drive safety bike, J.K. Starley's 'Rover'. And by the early 1890s the diamond frame more or less as it is today was in production, by manufacturers such as Thomas Humber.

This combination of steel tube and diamond frame has proved to be very enduring. It gives anyone setting out to design a better (rigid safety) bike a bit of a problem. For it makes the situation different from that of many other mechanical devices – cars, food-mixers and the like – where there are *numerous* design variations and opportunities for improvement. With the bicycle there is just *one* absolute and totally defined shape handed down to us by generations of frame-builders. And not only the shape, but *even the size of the tubes*, has been institutionalised.

Steel tubing – starting to get the right idea.

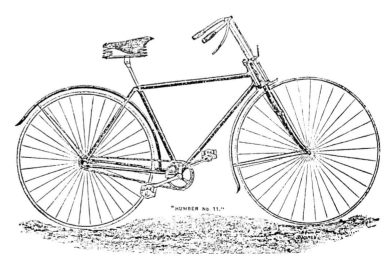

Alternatives to the diamond

The only real choice the would-be designer has is to pick a tube-set to suit the bike's purpose – touring, racing or whatever. To attempt anything else is to invite disaster – or at least, that is what history tells us. For although the 1890s' manufacturers such as Thomas Humber had clearly got it right, this has not stopped hundreds (if not thousands) of people from trying to change it. But in all this time, and despite all the effort, there have only been two variations that I would call successful.

The Moulton

One is Alex Moulton's small-wheel/suspension approach, using (firstly) monolithic cruciform and (later) multi-tube geodesic frame construction. Many subsequent small-wheel designs stem from Alex's 1960s original. That most of these derivatives are quite dreadful bicycles is not the fault of Dr Moulton or his design. It is still, if not a better bicycle, at least a viable alternative offering some real advantages over the traditional format.

(Above and left) The Moulton: not necessarily better, but certainly justifiable.

The Burrows Monocoque

The other and most recent alternative is my own moulded monocoque racing design, also now much (and equally badly) copied. Again, not a better bike (yet?) but offering the racing cyclist at least some advantage over 'iron sticks'.

With hindsight, it's easy to see why all the others failed. What they were trying to do was radically improve the bicycle, which of course is not possible. It is virtually perfect, so by definition any large change is bound to make it worse.

Tuning the diamond frame

The diamond frame can, however, be improved in small ways, and tuned to suit specific needs. So whilst it is undoubtedly the best way to construct a frame (provided the rider doesn't wear skirts!) there is still room for subtle variation. The nature of this variation will depend on the frame's intended use and the material of which the tubes are made.

Oversized tubes

For example, when the mountain bike first appeared it was seen as needing a much stronger frame than the traditional road bike, because of the extra loads that off-road riding would impose. So they started by abandoning the traditional $1\frac{1}{8}$" down tube and 1" top tube in favour first of $1\frac{1}{4}$ and $1\frac{1}{8}$", and then 1 x $1\frac{1}{2}$" and later some even larger diameters.

The Burrows monocoque:
not necessarily better, but faster.

Big tubes are beautiful – not just aesthetically but also structurally.

Needless to say, these early off-roaders were very heavy. It was not long before the usual pressures to produce lighter frames came into effect. Today, at least on top-end race models, there is little increase in weight over equivalent road frames. This has come about through improved tubing – meaning thinner walls and tougher alloys.

However, there is still a high degree of 'oversizing' of the tubes. This might seem logical but I personally do not think it is quite right. What a mountain bike needs is more strength and the logical way to add strength is by increasing wall thickness. Increasing the diameter adds stiffness more than strength, leading to a more 'brittle' frame. This is definitely not a good thing for off-road use. The stiffness required is more to do with the rider's power input rather than the terrain the bike is ridden over.

There will always be a minimum desirable frame stiffness needed to prevent shimmy. And the heavier wheels of the mountain bike, like the loaded panniers of the tourist or shopper, dictate a stiffer frame. But the best place to use oversize thin wall tubing would be for the track sprinter, who uses a lot of power over a short distance on a smooth surface. As he does this not very often, frame fatigue would be a minimal problem.

Aluminium

Changing from steel to aluminium is a different matter. The material's much lower modulus of stiffness means that it is essential to increase the diameter of the tubing, and in practice the wall thickness as well. Fortunately, aluminium being only one third the weight of the steel, the resulting tube is still both lighter and stiffer. This is

The bike, especially this one, is essentially a two-dimensional object.

due to a combination both of the properties and the nature of the mechanics of tubing. For each time you double the diameter of a tube, its weight doubles but its stiffness quadruples, whereas doubling the wall thickness simply doubles weight and stiffness.

This factor would give aluminium an enormous advantage over steel, except that aluminium, even when correctly welded and heat-treated (indeed, even if bonded) has a fatigue problem. This means that if the frame were allowed to flex as much as a steel one, it would crack sooner rather than later. For this reason all good aluminium frames tend to be over-built, making them even stiffer than basic function would require. This is why most aluminium frames have a rather unforgiving feel to them. They are, however, very efficient at transferring power to the wheels, especially for more powerful riders.

This additional stiffness does not, though, mean a harder ride, despite what you will have read in the cycling press. The reason for this is that the diamond frame is in effect a two dimensional structure. It has a height and a length, but its width is a function of the tube from which it is built. And it is the height-to-length ratio that has the greatest effect on the vertical compliance or comfort of a frame – not that it is ever very much.

The tubing size, wall thickness or even the material has the greatest effect on the torsional or twisting strength. This, unlike the vertical movement, can be noticed, especially when sprinting or riding a well-loaded tourer. The confusion in this area arises, I am sure, from the undeniable fact that our bodies are not well calibrated – unlike dial gauges and micrometers designed to measures things that we were not. It is not so much that we cannot detect anything. Rather, we tend to jumble up the various sensory inputs. And so a frame that is 'whippy' is assumed to be comfortable.

Torsional stiffness and vertical compliance

Balancing torsional stiffness and vertical compliance is one area of design where some fine tuning can be used. I started building 'baby' frames in 1982 for purely aerodynamic reasons. But I soon realised that they made a lot of sense for structural reasons also. And when the mountain bike started to evolve in this direction, this confirmed it. It might have stopped there, as I had by then started building monocoques. But, as I now have this rather nice job with Giant (who have not given up on the diamond just yet), the result is the compact road design. These frames should be more compliant vertically, although the difference will still be more imagined than felt. They should also have greater torsional stiffness for the same weight – not 'something for nothing' but a useful trade-off.

Greater specialisation

This type of fine tuning of the diamond frame will inevitably lead to a greater specialisation of design, with many designs becoming less universal. For example, the compact idea is less than ideal for touring, as the small rear triangle makes pannier mounting a problem. Other small changes can include ovalisation of tube in some areas. Even internal stiffening can sometimes be useful. But all of this is really just tinkering.

Fine tuning the diamond frame – even the UCI like the compact design now.

Beyond the diamond

Composite monocoques

If designers and engineers are ever to produce better bicycles, they have to forget about tubing of any sort. Instead they need to think about forming the frame directly, using some form of composite or even plastic material. This is getting very close to home, of course, as I started the ball rolling in 1982 with my first carbon fibre monocoque design. The reason was, as always on my racing bikes, to reduce the aerodynamic drag by producing the frame in one smooth form. Nonetheless, it was obvious there was tremendous potential for stronger and lighter frames than would ever be achieved with simple tubing.

One person who understood this even better than me was Chris Hornzee-Jones, a fellow recumbent builder. He went on to design an elegant monocoque mountain bike, briefly marketed as a Lotus. It was actually produced by Composite Air in California.

I like to think it was my original design that inspired Chris. But it was his understanding of composites that really made it work. The other pioneer of composite bikes was Brent Trimble in the USA. He produced some very nice moulded diamond frames that eventually became the Kestrel range.

There are now many copies around, either full monocoque or moulded modified diamond. Most seem to have been designed by people who lack Chris's understanding of composites, and who would probably be a lot happier working with steel tube.

To be fair, getting a monocoque right is, as I well know, not an easy thing. This is especially so with one intended for volume production. The first obstacle is the components. These have evolved over the last 100 years to suit the diamond frame and there is nothing in the Shimano (or any other) catalogue that was designed for anything other then the tubular frame.

Then there is the manufacturing process. You have a choice at present of 'slow and expensive' or 'slow and even more expensive'. Fine for the aerospace industry, but for the real world we need something a little more 'push-button'.

And finally, and even more important if the monocoque really is to become the superbike of the future, is the design itself. For whilst the simple aerodynamic forms I have been producing over the years are okay, to be able to reduce weight without sacrificing stiffness, a more scientific approach is needed. That means computers, of course, and Finite Element Analysis (FEA). Only using this approach can the variables of the frame shape, type of material, its thickness at any point and fibre direction be checked fully.

This is now standard procedure for many industries. Although FEA is not common in the cycling world, Trek did some pioneering work back in 1985 to help them optimise the tube sizes for their aluminium frame.

All this dictates that the monocoque needs a lot of money spent on it. This requires a large and profitable company – not a common thing in our industry. The ones that do exist tend to leave the real innovation to the smaller companies. Needless to say, I am working hard to try to change this.

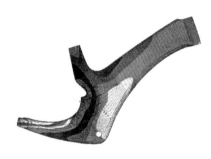

Useful tools, computers – my best friends all have them.

HPVs – recumbent alternatives

All of the foregoing refers to the rigid safety bike – still by far the most popular form of the bicycle, but far from the only one. Any designers who feel constrained by the diamond only need to move sideways a little to find the world of recumbents or Human Powered Vehicles (HPVs)[7]. As yet, no one is absolutely certain how many wheels HPVs

---

[7] *HPVs are not necessarily recumbents (or even wheeled vehicles) but most are.*

Laid back and cool, the Windcheetah trike has been evolving since the early 1980s.

should have, let alone what size they should be – or even where they should go! So there is still plenty of opportunity for using the tube in new and exciting ways, plus lots of scope for moulding.

This is an area that I have been working in for some time, and which I still find very exciting. I have tended to favour the monolithic or 'big fat tube' approach, but wouldn't claim it to be necessarily the best. It just suits my way of working.

Anyone with a special interest in HPVs should think about joining the British Human Power Club. Or if you think you might be interested, write for a copy of their 'how to do it' booklet. This covers the subject rather more fully than this largely 'upright' book can.

Dual-suspension MTBs

Also offering plenty of scope for new design approaches (and even the chance to make some money!) are dual-suspension mountain bikes. Not exactly brand new – Alex Moulton did it in 1988 – but they have suddenly come of age with the recent increase in downhill racing. Here many of the traditional drawbacks of suspension, like added weight and unwanted rider-induced bounce, are hardly a problem. This is because the riders are not so much pedalling as falling in a semi-controlled way down the rough side of a mountain. So the need is for 150mm of travel at both ends, some good brakes and sufficient durability to last at least a couple of races before bits start breaking off!

Many of the current designs are still based on the diamond frame, but the trend is towards more specialised designs. Many of them are driven more by marketing than functional concerns. But then we know more about marketing than dual-suspension, so why not?

City bikes

Moulded designs would seem a very logical way to go, except that things are changing rapidly, and the tooling can be very expensive and inflexible. However, my own RTM Moulded 'City' bike has shown that it can be done. I have been riding a prototype for two years now with no problems, I have not even looked at the chain. But there's still no easy way to mass produce it.

6. Aerodynamics

We have been using our wonderful two (and three) wheeled machines for competitive racing for 130 years so you might imagine that, as in other competitive sports, we would by now have discovered what the factors were that most affected their performance.

But, of course, you would be wrong. For as we all know, the first question asked about a shiny new racing cycle is, "How much does it weigh?" And from those not quite up with the latest 'Shimano', "How many gears does it have?" As if these were the defining factors in a cycle's performance!

True, it is nice to have a light cycle under you, especially when you are going uphill. And a good selection of gears is very useful, up or downhill. But these are *secondary* factors.

The importance of shape

What matters most, within reason, is not your bike's weight but *its shape* and *your shape on it*. For racing cyclists, or indeed anyone in a hurry on a bike, the single biggest problem is air. True, it is useful stuff for keeping cool (not to mention breathing!) but it really does put a brake on our performance.

To understand why this seemingly harmless substance is such a problem you have to go back to basics. I promised no algebra, but we will need a number or two, and the occasional diagram.

The Nature of Air

Firstly, air is a substance a bit like water. It has very similar properties to a liquid but is some 800 times thinner. And unlike a liquid it can be compressed, which can happen as objects travel through it. Air is a problem for two reasons. Firstly it sticks to things. Secondly, and more importantly for cyclists, it gets in the way.

The stickiness is due to the structure of the air which is made up of individual molecules. These in turn are made up of many atoms, which in turn… Anyway, there are these molecules which due to a phenomenon known as intermolecular attraction tend to stick to things. Not quite like glue, but well enough that the air flowing over any object flows not over the material of the object itself but over a layer of virtually stationary air molecules. These molecules in turn try to stick to each other.

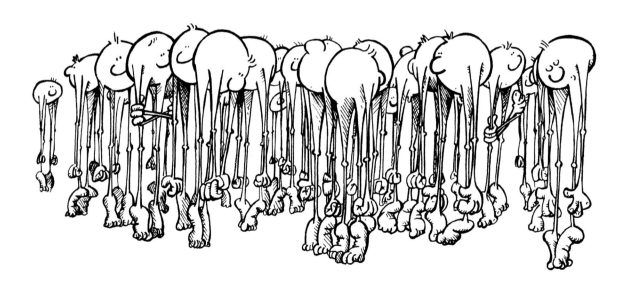

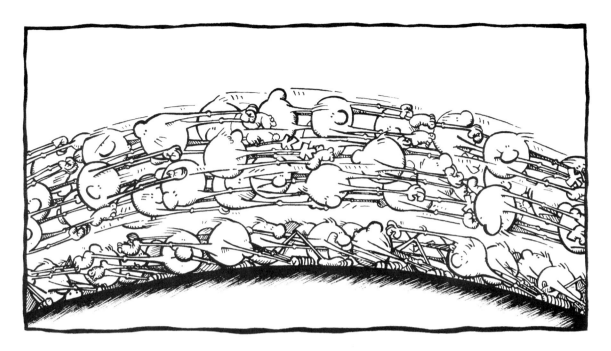

Air molecules stick to things. Other molecules flow over this stationary layer.

The boundary layer

The result of all this molecular interaction is a gradual change in the speed of the air flowing over the object. This is known as the boundary layer and can vary in thickness from one to one hundred millimetres – depending on several factors that I am not sure of and you really don't want to know about. The boundary layer can exist in two forms. Firstly as a 'laminar' flow, where each layer of molecules flows past in an orderly fashion. Secondly as a 'turbulent' flow, where there is a series of small swirls and vortices within the boundary layer itself, constantly mixing the streams of molecules.

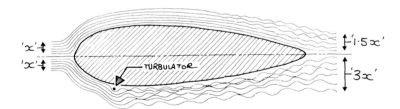

You can't see it, nobody can, and it won't make much difference to your performance, but disrupting a laminar airflow can sometimes lower the drag, for example, of fork blades or the so-called aero helmets. Thick insulation tape should do the job.

This boundary layer condition is something that concerns the people at Boeing an awful lot, but us cyclists only slightly. Both types of flow create friction, which is a real problem for things that travel fast. The space shuttle is a good example and has to be covered in ceramic tiles because the airflow causes it to glow red hot on re-entry. Fortunately for cyclists the problem is less extreme and regular Lycra is usually sufficient.

'Skin friction' is less of a problem for cyclists, not just because of our lower speeds, but because we are such awful shapes. Hence our big problem is simply the air molecules getting in our way, skin friction accounting for only 5% or so of total air resistance.

Playing CDs

This problem of shape is really to do with our ancestors, who were in all probability apes. They were well suited to swinging through the trees, but in aerodynamic terms were a bit of a mess. Unfortunate but understandable. (Equally unfortunate, but rather less understandable, is why our bicycles have evolved in the same way.) Birds or even fish would have been a much better origin for us, being as they are far better at avoiding air molecules due to their streamlined shape.

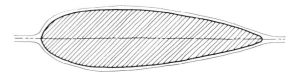

Pure theory, this one. It implies that your brake cables are creating more drag than, say, a fully aero down tube. Hard to believe, eh? That's because it's probably not true. It will be true in a wind tunnel; any length of round rod will have the same drag as a streamlined section ten times thicker. But this is only true in a clean airflow and on largish aeroplane-sized things. Aerodynamacists do not like talking about the air flow around small things; for small read 'bicycle sized'. I know this from my aeromodelling days when we had to do a lot of guessing. More importantly the bits on a bicycle are in a very messy airflow, the down tube is behind the wheel/fork and the cables are close enough to bars etc to confuse the issue. But this is a very nice image and you should commit it to memory for future reference.

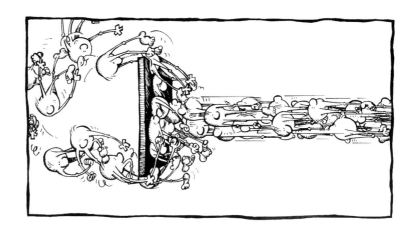

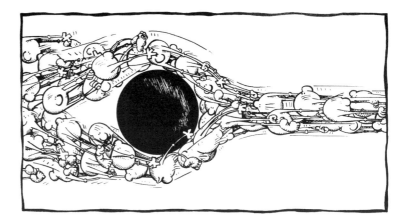

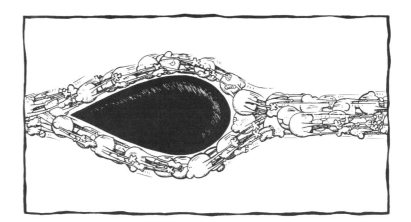

Give air molecules an easy ride!
Get rid of flat plates and use
teardrop shapes instead.

For shape is everything in the world of aerodynamics. And just how good a shape is at travelling through the air, as opposed to hitting it head on, can be defined as its coefficient of drag, or CD. The starting point for this scale of measurement is the 'flat plate'. This is a very convenient universal form that anyone working in a wind tunnel, where these things are tested, can easily use to set their instruments. It is also close to being the least aerodynamic shape possible and is given a CD factor of 1. That's it – 'one'! It does not matter how big the plate is, as this is not a measurement of the actual drag but the *efficiency of the shape*. So it is always 'one'. (Except when it is something else, as many things can affect the CD of even a flat plate. Again, you really don't want to know...)

What you *do* want to know is the CD of cycling's most popular shape: the 'rounded bluff form', which is the shape of most of not only our machines but also our bodies. This has a CD of 0.5 or half that of the flat plate. This is because the air can now squeeze past the front a lot easier, but will still create a large vortex behind the shape. So, a better shape but still not good.

Streamlining

What we really want is a streamlined form. The classic 'teardrop' or streamlined form has a nominal drag coefficient of 0.05. That's one tenth the drag coefficient of the round form for the same frontal area, which really is a nice thought. You can see why evolution was so hard on square birds! As with the other types of shape, this CD can vary due to other factors – one of which you *do* want to know about. This is the length-to-width ratio of the streamline. It needs to be about 4 to 1. Any longer will not really help in the cycling arena; any less can cause a sudden breakdown of flow, as with a round tube. Many so-called streamlined cycle parts (and in particular aero-frame tubing) have a ratio of 2 to 1 or so, and are therefore totally non-aerodynamic.

(Above) There's 10 miles of air molecules above your head. At altitude (above right), there's fewer, and so less drag.

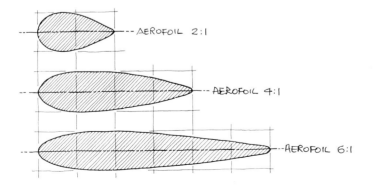

Not just the right shape but the right proportions. 2:1 is never aero, hardly better than round tube. 4:1 is best in most situations, 6:1 looks better but is only better in small sizes (less than 25mm long).

Air pressure and altitude

All of this effort to improve shape allows the molecules to slip past relatively unaffected, as opposed to building up in front as they would with a non-aero shape. This building up is, of course, a change in pressure. And it is this pressure that determines how many molecules there are in a given space. At sea level this pressure is about 100kPa or 14.5 pounds per square inch, and is due to the amount of air pressing down from above – 14.5 pounds of it on each square inch!

So one way of going faster is to find somewhere with fewer molecules to start with. Like up a mountain, where there is less air stacked up above, so fewer molecules to get in the way or stick to you. Not that the 'engine' will be very happy about this, as engines, both human and internal combustion, are quite partial to air molecules, especially when expected to run flat out. But even allowing for this, a racing cyclist will usually go faster at altitude, especially over short distances. For streamlined HPVs attempting the 200 metres record the advantage is thought to be around 1kph per 200m of altitude.

But trips to Mexico City are not really the answer, especially if you are planning to ride a time trial on your local dual-carriageway…

Interference drag

Before going on to suggest some possible solutions, one final and rather messy bit of theory for you.

Aerodynamics as a theory tends to be as simple as the object that you are designing. So if, for example, you are designing an airship, which is a very large simple form with very few significant bits on it, then all you need to do is make it the streamlined form as explained earlier and that's that. If, on the other hand, the object is an aeroplane, or even worse a bicycle (which is nothing *but* 'bits') then you have a real problem. This is 'interference drag' and is the result of things that are close to, or even worse joined onto, other things. It applies to objects that are streamlined,

(Below) Keep things apart! If a biplane's well streamlined wings do this to the air, what is happening behind your fork blades – not to mention your legs and frame tubes?

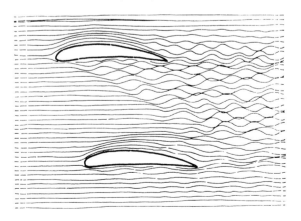

as well as the horrid-shaped bits that make up most of our bikes.

It is this theory that brought about the demise of the biplane, and it's why my bicycles tend to have monoblades. It's a big problem for aircraft designers. Much time is spent in the wind tunnel fine-tuning the bits, as the drag of a join can be greater than that of the bits themselves. Bike designers generally do not have access to this type of research, so we have to do a lot of guessing about this problem.

Putting theory into practice

So how can we apply all this theory? You have, in effect, four ways to reduce the drag of bike and rider. You could start by trying to reduce skin friction by unsticking some of the molecules that cause it. (Don't bother with Teflon™ coatings, as air molecules stick to everything the same.) This is an area of research that has occupied the best brains of the aircraft industry for a long time. For if they could achieve perfect laminar flow, they could more than halve the drag of a modern jet airliner! Ironically, whilst *they* are finding it impossible to achieve laminar flow, *you* as a cyclist can achieve it quite easily, as you are relatively small and slow.

Pre-turbulate

But as you are not streamlined as such, you don't actually want the airflow over you or your bicycle to be laminar. This is because, although the laminar flow does produce less skin friction as such, it is a very unstable flow and can break away from the surface totally for very little reason. Pre-turbulating the airflow to ensure a higher skin friction does in fact produce a lower overall drag, as the more stable turbulent flow will stick better around the curves of our lumpy bodies and bikes. This phenomenon was discovered by a Dr Prandtl in the 1930s but is usually referred to as the 'golf ball effect'. Yes, that's what the dimples are for. Dimpled balls go further than smooth ones.

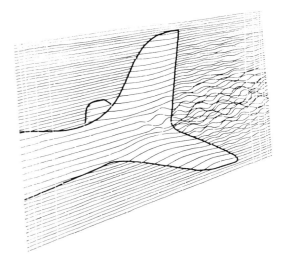

Even nice shapes meeting other nice shapes can be a problem. It can be made better but neither you nor I can afford the wind tunnel time.

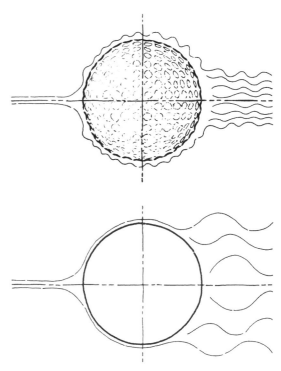

Four! Supposedly discovered when old balls went further than shiny new ones, this could represent the air flow over your fork blades. Pair of Penfold 700C anyone?

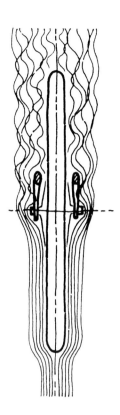

Yes, it really is this bad. Even with aero blades and a tri-spoke it is not good. One aerodynamic solution would be a monoblade with a 6" disc wheel; probably not a very workable solution outside of the wind tunnel, though.

In practice this will only affect your choice of clothing, so don't bother with the rubberised skin suit or even long sleeves. The texture of regular Lycra should be fine. But try to avoid wrinkles as they will cause extra drag. If you are very keen, and can keep a straight face, you could try shaving your legs but having a 'Mohican' stripe down the front to do the turbulating. You can forget about the airflow over most of the bike, as the air has been well churned up by the front wheel, and you can never achieve a laminar flow in 'dirty' air.

You could do another 'Mohican', though, this time using 6mm wide PVC insulating tape stuck down the front of your fork blades.

Reduce size

More useful is trying to reduce the size of things. For whatever the CD or theoretical efficiency of a shape, the smaller its frontal area, the less the actual drag it will produce. This is especially true of non-aero shapes like your body. Riding position is covered elsewhere but I must emphasise that achieving a good aerodynamic position on the bike can do more to improve your results than anything else (assuming you are the rider of a safety bike).

Making things smaller on the machine itself is rather more difficult, as the bits are already quite small and inclined to break as it is. Tyres are a good starting point, however. 18mm is the only thing to consider for serious riding against the clock. The good ones will always give a lower rolling resistance. But more importantly, the airflow onto and off of an aero rim is improved if the tyre is narrower than the rim. Leaving out a few spokes, especially in the front wheel, is a reasonable gamble on good roads or in the velodrome. But don't expect the wheels to last long.

Avoiding very oversize frame tubes will help a little, although they do produce a very light and stiff frame. One alternative is to use moderately oversize tubing and a micro frame design. This will give high torsional stiffness but lower frontal area, not unlike my new 'Compact' designs!

Improve shape

Far better than simply reducing the size of existing things is to make them a better shape. This gives bigger gains and lower strength losses. For example, if you halve the diameter of a round tube, you will halve its drag and its weight, but reduce its strength by a factor of four. Whereas changing the section of the tube to a streamlined form with a 4:1 ratio and the same frontal area will reduce its drag by a factor of ten. By adjusting the wall thickness it will also give a weight increase of 30% or so for the same strength.

Cyclists learned long ago that achieving a good aerodynamic position can do more to improve your results than anything else.

A selection of Mike's favourite bikes...

Von Drais's running machine.

Some of the aero bits you can buy...

❶ The seatpost is a good investment – 10/10.
❷ The chainset's just for pose value, really.
❸ The helmet gets 5/10 for aero usefulness.
❹ The bars are genuinely useful – 7/10.

(Above and right) Human-powered flight, demonstrated by the Air Glow.

High bicycle and Burrows monocoque.

Cinelli Laser.

(Above) Burrows Amsterdam. An experiment in town bike design, 1990.

(Left) A typical Dutch roadster.

Trek OCLV.

Giant TCR.

(Below) Mike's monocoque city bike.
(Right) An AM Speed, customised with some early tri-bars.

(Right) The Kingcycle road recumbent. No longer made, but a fine, fast machine. (Below) The Kingcycle Bean.

The Windcheetah Speedy.

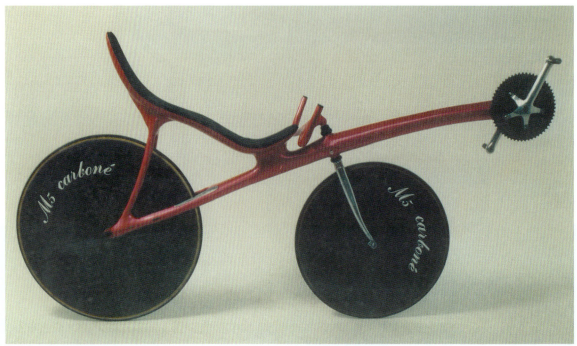

(Top) Chas Roberts road bike.
(Above) M5 carbon recumbent.

Burrows' Ratcatcher recumbent

Mike and his latest monocoque.

This is, of course, a theoretical situation (aren't they always?) and applies to single tubes in clean air. You can, however, get very good results from an aero-section seatpost, as the aero section is in the direction of the applied load. It is the one aero component that can actually be lighter than a non-aero one. Also, I have always felt that the airflow at this point was critical. For as we tend to scoop up the air with our bodies, there will be a high pressure area created. The gap between the legs is where it ends up. So not filling this area with a round seat pin and seat stays, etc., seems like a good idea.

Your forks can also be at least semi-aero, as they are in clean air. And, indeed, if you have a streamlined frame they are essential, as a pair of regular forks will create such a turbulent wake that the effects of an aero frame will be lost. But don't forget that the forks have an important role to play in pointing you in the desired direction. A really pure airfoil blade could easily compromise this function. Best of all would be to add a monoblade. It will seem heavy and look odd, but aerodynamically it has no equal for time trial or track use.

Aero-section tri-bars can also be useful, being up-front in the clean airflow – but not if they compromise your riding position. Aero cranks would be nice but nobody makes any yet. Wheels are covered elsewhere. Things like spare tubs/tubes, non-aero bottles, etc., should be 'hidden' behind you, well up above seatpin line.

Unfaired advantage

Finally, how to deal with that 'interference drag'? As I said earlier, a lot of time and money is spent by aircraft designers trying to solve this problem. But this is not an option for us penniless bike designers. This is a shame and leaves us with a big problem. For experience shows that, however good the shape of the individual tubes, the overall CD will be several times higher than simply adding up the bits.

There is, as I am sure you know, a fairy tale ending to this chapter. Not too long ago a handsome prince (well, all right, a scruffy engineer) discovered the answer: *the carbon fibre reinforced epoxy resin, cantilever, monocoque bicycle.*

I called it the 'Windcheetah', but most of the world knows it as the 'Lotus bike'. It solved the problem of interference by not having any! The 'fork' had only one side and everything (except cranks) was a true aero section. It was not especially light, the rear wheel was 16mm out of line with the front and the monoblade flexed very slightly sideways. So cornering on a rough surface is ... *interesting*. It was, and still is, the most streamlined unfaired cycle ever made – but I have finally worked out an even better one! You can see a picture of it on the front of this book.

7. The Wheel

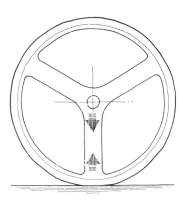

First used in Penny Farthing wheels, and now used for the Millennium Eye! That's right: the world's biggest wheel has individually-tensioned spokes, as that is the lightest, strongest possible design. But not the most aerodynamic. So whilst the compression loads in the tri-spoke are not good mechanical engineering, the compromise is acceptable for racing.

The wheels are the most important part of your bicycle – at least, that is what any racing cyclist will tell you. And with good reason, for the wheels are a large and crucial part of what a bicycle is. Indeed, around the turn of the century, the Americans referred to racing cycles simply as 'wheels'. But how much of wheel lore is science and how much myth? Do spoking patterns make a difference? And do disc wheels give a hard ride?

Evolution of the cycle wheel

The wheel started out just like the frame to which it was attached. It was built from wood. And it was built really quite well, for wheelwrights had been perfecting their skills for many years. But like the wooden frame it had its limitations. For not only was wood not the ideal material, it could only be used in *compression*. That is, where the spokes at the bottom hold up the hub.

With the arrival of steel, engineers soon realised that they could, as it were, 'hang' the hub from the top of the rim by using fine steel rods that could be tensioned individually. This meant not only could the wheel be adjusted to stay circular (not an easy thing with wooden structures) but that a large part of the structure was now in tension rather than compression – a much more efficient way of using material. Just imagine how much weight you could hang from a single spoke, then think of the sort of structure you would need to place below that same weight to support it.

The new form of construction meant that early cycle wheels quickly became very good. This was just as well, as the bicycle was evolving into the 'high wheeler'. In this design the front wheel was the largest part of the structure, its diameter being dictated by the rider's inside leg measurement. However, these early wheels had one drawback. For although they supported the rider's weight adequately, they did not transmit the power from pedal to rim so well.

This was because they were all built *radially*. That is, with the spoke going directly from the centre line of the hub to the rim opposite. However much tension you put in a radial spoke, there will always be a degree of 'wind-up' as you apply a twisting force.

The problem was eventually solved, as were many others, by James Starley. His solution was the tangential or crossed-spoke design. The wheel as we know it was with us. However, like the diamond frame, the detail has continued to improve. So we now have a selection of types from track racing to trade bike. And since the late 1980s, there has been a whole new area of development with the re-introduction of the 'solid' disc and semi-solid aero wheels. I say 're-introduction' as they are illustrated in Sharp's book, albeit in sheet steel! Much nonsense is talked about these new-style wheels – and not a little about the more traditional variety.

Traditional tension spoke

These are built with anything from 24 to 48[8] spokes and a shallow rim that can vary from heavy steel on the cheaper utility types to hollow rolled alloy on most of the better models. Spoking patterns vary from radial to cross-four. These non-aero designs are the strongest and lightest types of wheel, keeping as they do the amount

---
[8] *As few as 16 may be used in small wheels, eg. on racing Moultons.*

It's not quite science, but it's better than just riding around in circles for a bit...

Light and stiff – just add spokes.

of material in compression (that is, the rim) to a minimum and concentrating it in the tensioned spokes. This also produces a very tough wheel. It is far better at coping with abuse than a moulded wheel, which may appear to be very strong but will actually be relatively brittle.

I do not, though, believe that riders can feel the difference in ride quality even of these radically different types of wheel, let alone the difference that spoke patterns can make. I was once sent some wheels to review for a magazine; they had been built with 'wavy' spokes. The idea was that the waves would act like springs to give a more forgiving ride. Being an engineer I tested the wheels not by putting them on my bike but on a test rig. I loaded the rim to simulate a load some three times normal. Measuring this with a dial gauge and not with my backside showed a rim deflection of just 0.4mm. However, the 32mm cross-section tyres, that had been inflated to 500kPa (about 75 psi), had deflected *16mm*.

This did not seem unduly 'springy' for a wheel. I therefore checked a selection of others, from radial race wheels that had a deflection of 0.35mm to a old cross-three tourer that gave 0.5mm. The conclusion must be that deflections of this magnitude, let alone the small variations, are beyond the range of any part of the human anatomy to detect. This being so, the most logical spoked wheel is one with radial spokes – provided that it is not transmitting power or braking torque, and that you use an appropriate hub, as the radial spoke loads are high. This applies equally for track, road, touring and ATB wheels.

For the rear wheel, or if using hub brakes, a cross-three pattern would seem the best. Any more crossing and you start to lose lateral stiffness. Any less and there is still some wind-up. For racing wheels, where weight is important, there are some very nice 15/17 gauge spokes available. For bomb-proof touring, try to find some 13/14 gauge single-butted. These are expensive but virtually unbreakable.

Titanium should make sense for spokes but current ones do not seem to be very reliable. Aero-section steel spokes are another option, but these should really be used in conjunction with an aero rim.

Deep rim with tension spokes

Although closely related to the traditional wheel, this is the most recent style of wheel to emerge. There were aero-ish rims earlier, but they were no deeper than they were wide – and thus not very aero. The true deep rim was pioneered by Steve Hed in the USA, and HED Wheels still make some of the best available.

Deep rims are typically 30-75mm in depth, with 12 to 24 spokes, usually of aero or aero-ish cross section. Rims can be rolled or extruded aluminium, or moulded composite with or without a metal sub-rim – essential if you need to stop! These wheels will always have a small weight penalty compared to their shallow-rimmed cousins, but any difference in ride quality is, as I said, imagined. They may well feel different but this is due to the aerodynamics of the deep rim, which can be detected even in calm conditions. For this reason the *very* deep (60mm or more) rims are not a good idea for road racing, though the shallower ones are ideal. The deeper sections are perfect for time trial and track, being manageable in winds that would make a full tri-spoke a problem.

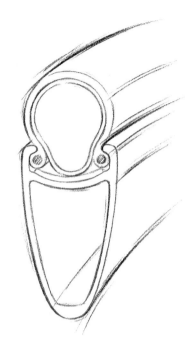

Moulded 3, 4 and 5 spoke

I find it curious that, although only very small numbers of bad bicycle frames are bought, vast numbers of quite awful wheels are sold to unsuspecting cyclists. One reason for this could be that cyclists do not know anything about aerodynamics, and the people who produce the wheels are equally ignorant about the mysteries of airflow – not to mention basic engineering.

The original tri-spoke and five-armed aero-spoke were good examples of a lack of even the most basic understanding of the problems involved. The rim was hardly deep enough to be streamlined and the spokes came straight from a wheelbarrow. The best example of a good wheel is the Specialized (now HED), which was designed and built by Du Pont and uses proper aerofoil sections. It is also beautifully moulded using RTM technology. Wheels like the Specialized and the new Mavic are, in still air, probably the most streamlined available. If conditions are not too windy, they are also satisfactory on the front for time trial use.

My current least favourite wheel is the Spinergy Rev-X design, which despite being a best seller is ugly and non-aerodynamic – because the wheel has a biplane rather than a monoblade structure. A number of the early wheels also failed catastrophically.

Discs

(Above and left) Deep section rims, tri spokes and discs: not so light, but fast.

The original aero wheel and still the best – or at least, the most *aerodynamic*, if well done. However, mostly they are not so good, being flat sandwiches of foam and composite around 20mm thick. This is not very good structurally and, although better than a spoked wheel, not very aerodynamic.

Better for both reasons is a conical design. The extra width at the hub adds stiffness and improves the aerodynamic shape. It also hides the hub, a not very aero bit that we tend to forget about. Most of these designs are still moulded compression structures, but at least one design (I think from Russia) uses an Aramid skin under tension. Very neat!

Best of all aerodynamically is the lenticular or curved side design. But again, it has to be done properly. The only good one I have seen is by FES of Germany. Even that had a rather narrow section rim. It is always better for the smooth flow of air if the rim is at least as wide as, or even a shade wider than, the tyre.

Many organisations now allow you to use add-on discs, either clipped or bonded to conventional wheels. This is just as aero as the more expensive alternative and has been the norm with HPV builders for years. It is all I have ever used.

Disc wheels are for use only on the rear, except on indoor tracks, as the effect on the handling of a bicycle can be quite dramatically negative. Contrary to popular belief, though, disc wheels are even better (on the back) in a cross-wind.

For off-road use the regular shallow rim design works well. Deep rims are particularly pointless, as there is no airflow behind a 2"

knobbly tyre! And the extra rim stiffness is of no advantage. I would like to see the opposite – an extra shallow, tough but flexible rim, maybe of composites, that would absorb major impacts. This should reduce rim, frame and tyre damage. For the same reason moulded wheels would seem to be especially silly for off-road use.

Recumbents can generally use a regular wheel with 'added' discs. As the wheels are usually smaller you may even be able to run a disc on the front. Three-wheelers can create a lot of side force, which is not something cycle wheels are designed for. Therefore keep the diameter down to 20" maximum for the main pair of wheels. A trailing single rear wheel is less heavily loaded and I have found 700C fine here.

How big?

It has been the way of wheeled vehicles of all sorts that, as they evolve, the wheels get smaller. This is not something that is seen in the bicycle, but then it effectively stopped evolving 100 years ago. So should we give evolution a nudge? Do we need smaller wheels? Or maybe even bigger ones? After all, Francesco Moser used a one-metre diameter rear wheel for one of his record rides.

To be honest, the true answer is 'maybe'. The correct size for any given wheel is very hard to calculate, as there are many – often conflicting – factors.

First and most obvious is the rolling resistance. Theory tells us that, as the diameter of a wheel increases, its resistance to movement decreases proportionately. That is, if you double the diameter of a wheel, you will halve the effort required to roll it along a smooth surface. And equally, if you halve the diameter, you will double its resistance.

The problem with this theory is that it assumes a perfectly smooth surface – something that is sadly lacking in the real world. Even the best velodrome track is a series of small waves and ripples. All of which means that resistance increases even more than basic theory would suggest as wheels get smaller. It is worth remembering that those other vehicles that were evolving their smaller wheels were also evolving rather more powerful engines – not an option for us single-cylinder folk.

The resistance to movement that the basic theory measures is caused by the difference in the energy required to bend the material of the tyre and tube where it meets the road, and the energy returned as the material springs back into shape. This will be affected by:

- how easy the material is to bend,
- how quickly it returns to its original shape,
- how much material is involved, *which can mean the thickness of the tyre tread or how large the 'flat bit' is*, which latter is a function of:
- tyre pressure, *and*
- how far the material is bent, *which is a function of both*:
- wheel diameter, *and*
- tyre cross-section.

There is also the stickiness of the rubber, but for cycling the rubber compounds are all so hard that it can be assumed to be minimal and constant for all tyres. (More on tyres in the chapter about them.)

As mentioned before, the other factor that the basic theory does not take into account is all those bumps and ripples in the road that we are constantly having to ride up and down. They are like a series of hills in miniature, and have much the same effect on performance. We climb hills slowly and descend them quickly because of the effect of gravity. If we were able to do this on the Moon, or even in a vacuum on Earth, this would get back all the energy we expended in climbing the hill by the time we got to the bottom. But in the real world the otherwise very useful air rather spoils things. Specifically, the air resistance on the fast descent rises faster than it decreases on the slow ascent. (More on this in the chapter on aerodynamics.) This effect, although of course much reduced, occurs every time our wheels go up and down a bump. It affects all sizes of wheel. But for a smaller diameter wheel, the bump is relatively

> The best suspension in the world, a 50mm mountain bike tyre will 'eat' a 25mm stick, and all that happens is a few grams are bent out of shape momentarily and then spring back thanks to the air pressure returning most of the energy expended. Contrast this with a solid tyred rigid frame, which would have to 'climb' the stick raising both cycle and rider upwards the momentum of which would mean not rolling down the other side of the stick thus losing all energy expended and coming down with another bump. Slow and uncomfortable. Thank you again, Mr Dunlop.

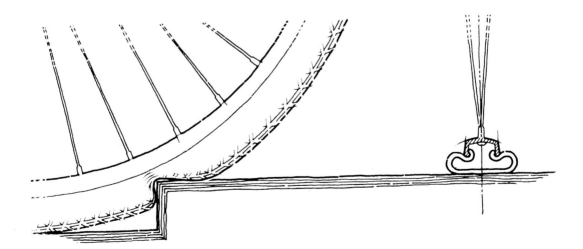

larger, so the slowing up is more pronounced. Even worse, small wheels are more inclined to 'fly' over the bumps. That is, not to roll down the other side, thus losing all the energy used in reaching the top.

The effects of road surface can also be varied by changing tyre pressure. A softer tyre will be more absorbent of small bumps and less inclined to fly. It will, however, have a higher rolling resistance due to increased material flex. So there is a balance to be found. From experience, for pure performance the balance should be tipped towards hard tyres unless the road is very rough.

Rolling resistance, however, is not the only factor that concerns cyclists. If you are interested in racing, the wheel's aerodynamic drag will be a very important consideration. There is therefore a certain logic in saying that to halve the wheel's diameter (although more than doubling the rolling resistance) will reduce air resistance even more. This is because the air resistance at racing speeds involves some 90% of the rider's energy expenditure whereas rolling resistance accounts for around only 7%. This leaves just 3% or so to cover all the mechanical losses due to chain and bearing drag.

Another factor to consider is wheel strength: smaller is always stronger. Also, there is convenience of use. This is not so much a problem for racers, but for commuters a small wheel allows for a smaller, neater bike to be designed around it.

Possibly the most important consideration is availability. Not many of us have access to the kind of resources that Francesco Moser used. So however much you might like a one-metre wheel, 700mm is the biggest you are likely to get!

Finally, accelerating the mass of a wheel famously requires twice the energy as accelerating a similar amount of weight fixed to the frame. This is because you have to accelerate it around the wheel and also *forward*. So small wheels accelerate faster. Trouble is, it is such a small number that it is almost impossible to measure, so don't let it worry you.

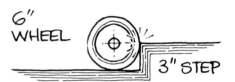

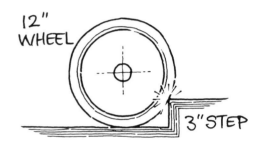

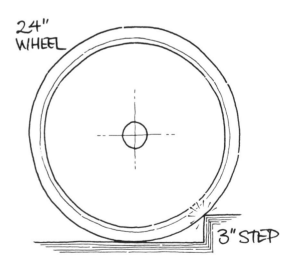

In the real world, a 6" wheel could have a theoretical rolling resistance of less than 1% but as the illustration shows, this can easily rise to 100%.

Getting the balance right

Let's start with the nearest to the ideal environment, the velodrome. Many riders are now using 650C wheels, especially on the front in place of the regular 700C. As I explained earlier, there is a good argument for this, especially as speeds get higher.

The problem is that the designers of the bikes are not taking advantage of the smaller wheel to improve the frame's aerodynamics. Indeed, it is often the opposite and you see an extra 25mm of unstreamlined head tube replacing the 25mm of relatively streamlined tri-spoke. Also, you do not halve the drag of a wheel by halving its diameter. A large part of the drag of an aero wheel comes from the hub and the attachment point, which is constant for all diameters. And even if done well, I think that for the 4000 metre pursuit, where it is most often seen, the advantage still lies with the 700C.

There would be a better argument for using 650C for the 1000 metre time trial, where speeds are higher and the small acceleration advantage is of more significance. One area in which I definitely think 650C would be an advantage is the 200 metre sprint. Here the speeds are the highest of any event and the lateral stiffness would be a useful bonus, likewise the stiffer frame.

For time trial and triathlon the balance is still firmly in favour of 700C. In these cases speeds are lower and surfaces rougher. Also, there is a much greater choice of wheels and tyres. I know a lot of triathletes are using 650C front and rear, but they are wrong.

Road racers have no real choice, as they have to be able to change wheels anytime.

For touring I still favour 700C, or even our old British favourite 27" is nice if you can get them. Although the 26" ATB size gives a very tough wheel and an extremely good choice of tyres, at the speed a tourist travels rolling resistance is a significant factor. Over the long distance typically involved a lot of energy can be saved (or miles added) by using the larger diameter wheel.

For city use I feel the balance is tipped the other way. My firm favourite for the urban environment is therefore the 26". It is convenient, tough and there is a good choice of high performance semi-slick tyres available, ideal for all round use.

For the mountain bike itself, the 26" is the only choice. For although the 700C with a fattish tyre will give a measurably lower rolling resistance (because off-road the difference in wheel diameter is even more apparent), its large size makes the bike unwieldy. Equally inappropriate would be smaller wheels, although I have heard that some people are using 24" diameter as a way of getting more suspension travel on downhill bikes. This does not sound like a good idea, but as the downhill world has almost reached the 4" cross-section tyre a 24" wheel is needed to keep the overall diameter the same.

Suspension and small wheels do actually go together very well, but for city bikes. The suspension allows the small wheel to ride up and down the bumps just like a large wheel. Alex Moulton did this back in 1963 to good effect.

Recumbents tend to use smaller wheels for reasons of aerodynamics and design convenience. But designers should not forget that if they can improve the aerodynamics to the point that they can double the speed of a regular cycle, then they will have doubled the rolling resistance as a percentage. This assumes that they are still using 700C wheels, which is not very likely. So in reality rolling resistance could be as high as 25%, which is well worth trying to reduce.

For the more compact rider...

I realise that this chapter is rather sizist, assuming as it does that we are all of average height – which, of course, we are not. For big boys and girls there is not much I can suggest, as there is little choice in the larger sizes. But for the more compact lads and lasses we can do a bit.

The 650C is a good replacement for the 700C for road race and touring cycles – as is the 26" ATB wheel. Very small (14") ATBs could use 24" wheels to maintain good balance and proportions. Sadly, few of the big manufacturers have done anything about this yet. I am trying to convince my lot what a good idea it is. But for the moment it will probably mean a custom-built frame if you like the idea.

8. Tyres – Annular Pneumatic Suspension

I have added this extra bit on tyres because, although they are part of the wheel, they are a variable. Also, some factors, such as tread effect, are a separate issue.

The pneumatic tyre is without doubt one of the cleverest and best things on a bicycle. I am sure it is one of the key factors that made the bicycle a universal product. Without Mr Dunlop (or if you prefer, Mr Thomson 43 years earlier[9]) we would still have the bicycle – but only as a toy.

Solid tyres

Unfortunately the merit of pneumatic tyres is often not appreciated by 'designers' who attempt to 'improve' our simple steeds. So one of the first things to go are the 'pump-up' tyres, to be replaced by 'high tech' solid ones. This they do because, as they see it, getting a puncture is so awful they would rather be abducted by aliens. (We can only hope!)

A more balanced approach is taken by the average club rider – someone who actually *uses* a bicycle. He is quite happy with pneumatic tyres and simply carries a couple of spare inner tubes in his back pocket. This, combined with quick-release wheels, makes a puncture little worse than tea with the mother-in-law.

However, life being what it is, some of these punctures do seem to occur at the most inconvenient times and places. The idea of eliminating them completely can be very tempting. So just in case *you* are tempted, this is what the solid tyre will do for you...

Firstly it will take away something you probably did not know you had: your suspension. That's right – *your suspension*! You thought only mountain bikes and Moultons had suspension. But you were wrong. Any rigid bike can have anything up to 45mm of travel at each end. And very efficient suspension as well. It not only gives the

---

[9] *Thomson's tyre was not used on a bicycle but could have been adapted.*

rider an easy time of it, it also does an excellent job of isolating the frame from high frequency vibrations that would dramatically reduce its life expectancy. Yours would also be reduced. But that would be because you would probably give up cycling if you had to use solid tyres!

If you did *not* give up, you would actually get even fitter, as you would be working a lot harder. This is because the rolling resistance increases dramatically with all solid tyres. (The exception is steel on steel, but that's only for train drivers.) The reason is twofold: the amount of material being deformed as you roll along is increased; and its spring-back is much slower, not having high pressure air to push it back.

With solid tyres the amount of grip you get will generally be a lot less. This is because the contact patch is usually smaller and the material less grippy. Solid tyres are also very heavy – being solid, they would be. Adding this much weight to the rim of a cycle wheel makes even the lightest of bikes feel totally dead.

So resign yourself to the inevitable and carry a spare inner tube. You can even employ a philosophical approach, realising that you can only have good days by having some bad ones. So, if you are really lucky next time you get a 'flat', it will be raining, and your spare will already have a puncture…

Racing tyres

Back in the real world you still have a few choices. Traditionally all racing was done on 'tubular' tyres glued on to 'sprint' rims. These still have a lot to commend them. For velodrome use they should still be everyone's first choice. That is, unless you are just starting out and on a tight budget.

For time trial or road race their advantages are less significant and the disadvantages more so, so you will have to make a choice. For riders likely to be in the prize money, tubulars have a couple of big advantages for time trialling. Should you get a puncture, they can be changed in little over a minute, which could keep you in the money! Contrast this with a high pressure tyre and inner tube, which would typically take more than three minutes to change. And for road racing 'tubs' can be ridden on even when flat. You might well ruin the tyre and rim, but that's racing. With HPs there is no choice – you have to stop.

For the rest of us who are unlikely to see the chequered flag there is little reason for the extra expense of tubs and the mess of gluing them on, although they do allow a very light wheel to be built. The benefit might be a difficult thing to measure but it does feel very nice. And the best of the 18-20mm narrows show a measurably lower rolling resistance than even the best HPs. So maybe if you are an old tester like me, still hoping for a personal best, you could treat yourself to a pair of Sunday-best wheels.

But for the bulk of your riding, tubs are not worth it. The modern HP tyre is very good. Rolling resistance is better than many 'normal' tubs and they are a lot less hassle. My own favourite is still the Michelin Hi-lite. It has been ever since I tested a whole pile of tyres for rolling resistance. It is almost as good as anything you can buy, but is still a reasonable weight and not likely to puncture too easily.

DIY testing

If you would like to check rolling resistance for yourself, it's not difficult. You will need to find a suitable road, preferably one with not too good a surface. It should have a gentle slope to roll down; gravity is a very consistent power source. Do not exceed about 15-20 kph (9-12 mph) or the aerodynamic variations will mask the tyre differences.

You should try to roll for about 30 seconds. At this distance you will need to test each tyre at least six times to get an accurate result. A shorter rolldown will mean more testing.

Mark out a starting point on the hill with chalk,

Testing tyres can be hard work.

Roll Down Tests on ATB Tyres

Mike Burrows 1996

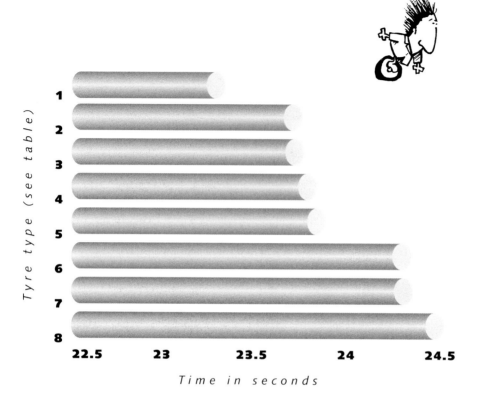

	TYRE TYPE (all by Michelin)	PRESSURE (atm)	WEIGHT (g)	SECTION (mm)	TIME (secs)
1	SLICK 2"	4.0 both wheels	560	45	23.24
2	TAIGA 1.95"	4.0 both wheels	610	43	23.66
3	JOGGER 1.90"	4.0 both wheels	590	40	23.67
4	SAVANNE 1.95"	4.0 both wheels	460	41	23.74
5	SURMOUNT 1.75"	4.0 both wheels	630	38	23.79
6	WILD GRIPPER LITES (L) 2.1" front, 1.95" rear	3.5 front, 4 rear	590 f, 580 r	47 f, 43 r	24.25
7	WILD GRIPPER LITES (Comp) 2.1" front, 1.95" rear	3.5 front, 4 rear	650 f, 600 r	47 f, 43 r	24.26
8	WILD GRIPPER LITES (Comp) 2.1" front, 1.95" rear	2.75 front, 3 rear	650 f, 600 r	47 f, 43 r	24.43

Not everything you need to know about rolling resistance, but it's a start.

and another some six metres down the slope. Assuming you are doing this test alone, you need a stop-watch that you can operate on the move. You start rolling at the first mark, and clock on as you pass the second mark. By that time you should be in a comfortable riding position that you can repeat.

The final mark, where you clock off, needs to be very clear, as you will be travelling at some speed by then. Ideally you need a road that runs up the other side, so you can be slowing down at this point.

You will need a very calm day (or more likely night) for the test. Take a track pump with gauge to keep the pressures constant. With this you can also check the effect of tyre pressure changes. You will probably discover, as I did, that anything over about 7atm (700 kPa or approx. 100psi) simply gives a hard ride.

The more roll-downs you do, the smaller variations you can detect. But you do end up riding back up the hill an awful lot.

This test method may not sound very scientific, but it can give very accurate results. Be warned, though: it is no use trying to test variations in the aerodynamic drag of cycles or clothing in this way. The variations on even the stillest night are greater than the variations you need to look for.

Here is one suggested solution to this problem. Use two riders of similar size and weight on similar bikes which have been 'tuned' to roll down the hill at the same speed, side by side. You can then change something on one rider and see if it makes a difference. I have not yet tried this system, but it sounds good.

Off-road tyres

Off-road gets much more interesting, as rolling resistance is only one small factor. And even that does not always make sense. For example, a narrow tyre will generally have a lower rolling resistance than a larger cross-section of similar construction. This is because you can get more pressure into the smaller tyre, so less flat bit on the bottom.

But it is different when your tyres are all on the fat side. They will all take more pressure than you would ever want to use. So the larger cross-section can have a lower rolling resistance on *any* surface. And on something like grass, it will be even better, as it sits on top, rather than digging in.

However, rolling resistance is not normally of

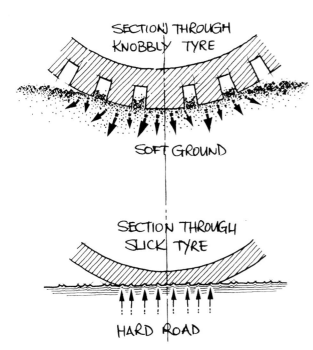

Which is softest? If the surface you are riding on is softer than your tyres (i.e. mud) then your tyres can stick into it and a tread is useful. If the surface is harder than your tyres (i.e. concrete) then it will have to stick into your tyres and so a lack of tread will give more grip as there is more rubber for the texture of the road to grip. The real world being an erratic mix of these and many other surfaces, some sort of compromise is usual for most riding.

great interest to off-roaders. They are much more interested in grip, which means tread design. You might think this would also be of interest to roadies, for they have to negotiate the odd corner. It is, but all they need are slicks. Nice smooth rubber, dry or wet, always gives best performance and grip. For when you are riding on a surface that is harder than the tyre, it is the texture of the surface that sticks into the tyre. Having a tread pattern simply takes away some of the rubber that would otherwise be gripping the road. You could, of course, aquaplane with slicks, but you would need to be doing around 200kph (125mph)!

Off-road you do need tread, at least most of the time. This is because the surface you are riding on is often softer than the tyre. The problem about trying to be scientific about off-road is the variation in the surface that has to be coped with.

For pure performance a slick tyre would be ideal but not very versatile. The current trend is towards a *near* slick, with large side knobbles to give a bit of security when laid over in softish conditions but not absorbing power when climbing. For that is the problem with large, aggressive, knobbly tyres. They might do quite well in a rolling resistance test, but when you start putting power through them the knobbles bend back, absorbing some of your energy. Knowing when you need to put on the knobblies is down to experience and individual skill.

In wet and muddy conditions there is no choice. Narrow (1.5") open tread knobblies are the only thing. Less mud sticks to them, so there is less added weight and they are less likely to stick in the frame. The trade-off is that the smaller cross-section is more likely to bottom out and give you a snake bite puncture on tree roots and the like.

There is at the moment an enormous variety of tread pattern and tyre colour, all promising to do this or that. A lot of people think that this is mostly a marketing thing. It is not. It is *totally* a marketing thing!

Tread design, I am sure, is a bit like aerodynamics but in reverse. What is bad at slipping through the air is good at not slipping sideways through the mud. And this shape is, of course, the square. By a remarkable coincidence this is exactly the shape of the tread on off-road motorcycles. It's not that the motorcycling world does not use hype to sell its products. It's just that the tyre is too important to them to risk! So for my money all our tread patterns should be based on the square, with the spacing and depth of knobbles to suit the conditions.

Another thought. Forgetting about wear for the moment, there seems little point in having the tread stick up any more than it will stick into the ground. From experience this will be around 3-4mm on a typical dry course.

There are, though, many conditions that defy analysis. At the end of the day, there is a lot of security in a pair of big fat knobblies.

Street and touring tyres

Not quite so daunting is choosing the right street or touring tyre. Pure slicks are an option and will give best performance and grip, especially on wet city streets. But they are a bit fragile and you cannot easily tell how worn they are until it is too late. Also, if you should venture off road even a *little* bit in the wet, the grip will be close to zero!

Far better to choose from one of the many semi-slicks with a minimal cut in the tread. This lets you know how worn they are and gives you just a little security on the odd muddy towpath. And don't choose too narrow a cross-section. Most of the new generation of street tyres have a good quality light carcass, and a big fat tyre gives a lot of security. Needless to say, big fat *cheap* tyres are *not* nice, so don't buy them.

A lot has been claimed for some of the 'special' inner tubes, which are usually latex. I have never tested any for rolling resistance, although they do look good. So if you have the money you could treat yourself.

9. Transmission — Gears Without Tears

*H*aving lots of gears is very fashionable, but they add weight and complexity. Do we really need them?

Certainly if a vehicle's power source works well under all normal combinations of load and speed (like steam engines) there is no need for variable gears. But if the engine works best over a relatively *restricted* output range (like petrol engines and most cyclists) gears can be a great boon.

Back in 1901 Henry Sturmey, a great champion of the bicycle but also founding editor of *Autocar* magazine, put it this way:

"…the human being is not unlike the internal combustion motor. It has been found absolutely indispensable for the latter that variable speed gears should be provided, simply because the engine cannot develop more than a given amount of power, and if its speed is reduced unduly owing to the heaviness of the drive, its power declines so rapidly that it cannot take the ascent, and although the human motor possesses far more elasticity than the petrol engine, its limits are very soon reached, and a variable speed gear is the correct thing for the rider who desires to combine the maximum of distance and speed with the minimum of exertion."

We therefore need gears to help match our limited power output to widely varying load conditions. Then we can cope better with rough surfaces and smooth, ascents and descents, headwinds and following winds, load-carrying and travelling light, racing or just ambling along.

Primary drive

Before looking at variable gearing itself, we need to take a quick look at the *primary drive train*. In other words, how the rider's power gets delivered to the back wheel.

Most early bicycles had the pedal cranks connected directly to the front wheel. The ultimate form of the front drive was the graceful and very efficient Ordinary or High Bicycle, often known by its derogatory Victorian slang name Penny Farthing.

Step-up drive – and how we measure it

The development of the bicycle chain enabled use of smaller driving wheels. Drive was fed to the *rear* wheel from cranks mounted on the frame. By using a relatively large chainwheel on the crank shaft and a much smaller one on the rear hub, the drive was stepped up. The rider could therefore travel much further for each turn of the pedals.

The step-up gear was expressed in terms cyclists of the time could understand – the equivalent Ordinary front wheel diameter. So, a 48-tooth chainwheel driving a 24-tooth rear sprocket on a 28" wheel gave a gear of 48/24 x 28" = 56". This system is still used in the English-speaking world.

An alternative system originated in France, *le dèvelopment*. This is the distance in metres the bicycle travels for one turn of the pedals. You can convert the British figure to the metric pretty accurately by multiplying by 0.08. Conversely a development figure can be converted to the British system by multiplying by 12.5.

Alternatives to the chain

For a 100 years the vast majority of bikes have used a chain step-up drive. It is a phenomenally reliable and efficient system, with losses as low as 1% at high power input levels[10]. But there are other step-up drive options, the main ones being belt drive and shaft drive.

Belt drive

Flat belt drives considerably pre-date the bicycle and are mentioned in the earliest cycle gearing patent (1869). However, they are not very practical for bikes because of their tendency to slip, especially in the wet or if they get oily. A *toothed* belt can overcome such problems.

Modern toothed belts are much more durable and efficient than they once were. Materials such as urethane or Kevlar™ are used. Millions upon millions of car camshafts are driven by toothed belts, where chains were once the standard solution. But car engines have plenty of power to spare and the slight loss of efficiency through using belt drive is much less significant than it is to a cyclist. The inefficiency stems mostly from the belt having to bend round the front and rear sprockets. This typically absorbs more energy than the pivoting of a chain over a chainwheel and rear sprocket. There may also be some loss through slight stretching of the belt under load.

Low maintenance, weather resistance and cleanliness have all

---

[10] *As early as January 1930 The National Physical Laboratory at Teddington measured the efficiency of a Renold bush roller chain drive as being between 98.1 and 98.9%.*

been arguments for the belt. During the last 20 years toothed belts have occasionally been used on production cycles, especially short-range folders and commuter bikes. In the mid-1980s three folders used belt drive – the Japanese Bridgestone Picnica, the Swedish MicroBike, and Mark Sanders' Strida (now back in production as the improved Strida 2).

Apart from the efficiency issue, belt drives have other disadvantages. They cannot be used with derailleur gears, hub gearing being the only practicable multi-speed option. Belt sprockets, being wider than chainwheels, are of necessity heavier. And unlike chains, the length of a drive belt cannot be varied to suit a particular machine. But most awkward of all, the belt cannot be split to pass it across the chainstay – a big problem with any frame incorporating a rear triangle. At least one designer keen to use belt drive had to incorporate a removable section of heavily stressed seat stay to enable the belt to be fitted and serviced. Expensive and structurally undesirable.

Belt drive therefore has a small niche in cycle design but does not show any signs of major growth at present. For most purposes the disadvantages outweigh the advantages.

Belt drive: better on a packaging machine than on a bicycle.

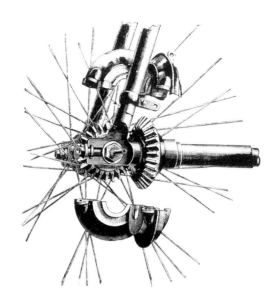

Shaft drive: again, better on a milling machine than a bicycle.

Shaft drive

Shaft drive has been around for well over a century. At first sight it seems like a very good idea. A nice, compact bevel gear instead of a chainwheel driving a shaft with a smaller bevel gear on each end, which feeds power to a final bevel gear on the hub. Usually the whole lot is enclosed to protect it from the weather. No oily chain to mark your clothes, fall off or wear out.

However, there are disadvantages to set against this. Transferring the drive from the cranks to the rear wheel via two sets of bevel gears is likely to lose about 2% of input power for each set of mating bevel gears and more for the bearings supporting the drive shaft – perhaps a total of 6% as compared with nearer 1% for a chain drive. And this ignores any torsional losses through the drive shaft twisting under load. This is a wind-up in more ways than one, and can make a poorly-designed shaft-drive bike very unpleasant to ride.

Obviously derailleur gearing cannot be used – no chain, no derailleur. So, as with belt drive, the only mainstream variable speed option is hub gearing.

But the biggest difficulty of all is the difficulty of adapting a standard bicycle frame to accommodate shaft drive. The drive shaft clashes with the traditional chainstay, so a retro-fit is impossible. The usual solution is to modify the chainstay so that the shaft actually runs *inside* it.

So, shaft drive is another solution with a small but limited niche. Every few years someone, somewhere in the world, launches a shaft-drive bike. But they never make much impact. It's an expensive way of building a less efficient bike with limited gearing options, and which may be less pleasant to ride.

The virtues of single-speed

Having eliminated belt drive and shaft drives, we are left with the good old chain. It's staggeringly efficient, inexpensive, proven and adaptable. But before looking at the multi-speed options it supports, let's look briefly at single-speed chain transmission.

Why? Because it's light, cheap, reliable and as efficient as it possibly could be. It's also used by the majority of cyclists worldwide – people in third world and developing countries who use their bikes day in, day out as primary transport and for load carrying. It's cost-effective and it works.

Single-speed transmissions can be fixed-wheel or free-wheel. We are all used to being able to stop pedalling and freewheel along. Virtually all bikes today let you do this. This is because of a major breakthrough in Victorian cycle technology that we all take for granted: the automatic freewheel.

But with single-speed chain drive you can, if you wish, turn the clock back 100 years and have a *fixed*-wheel. In this case the rear sprocket is rigidly screwed to the hub, so that the pedals must revolve whenever the wheels do. Some experienced cyclists prefer this, because it gives them more control, particularly on winter roads. It also enables them legally to dispense with a rear brake (provided the sprocket has a locking ring fitted), as they can slow the rear wheel by back-force on the pedals.

Single-speed fixed-wheel is the standard for most track racing. It can also be a great boon for some disabled riders. Special fixed-wheel conversions of hub gears and derailleurs have sometimes been made for such riders.

Cyclists who are torn between fixed and free-wheel can buy a Fixed-free Drive, which enables a choice at the click of a trigger.

Another once popular option is a double-sided or reversible hub. This has a different sized sprocket on each side. To change from one to the other, the rider has to stop, take out the wheel, turn it over and reinstall it. The sprockets can be fixed or free-wheel, or one of each. At the expense of oily hands and some inconvenience they give the rider a choice of two speeds or drive modes while adding little weight and no efficiency losses. But does anyone today make such hubs?

Apart from track racing use, single-speed transmission today is almost always combined with a freewheel. Standard uses on production bikes are for very young children's machines, basic roadsters, and lightweight compact folders such as the Cresswell Micro Lite. But occasionally individualists go against current fashion for multiple gears and retro-fit single-speed. There are even a few mountain bikers who have decided that a low-geared single-speed (say 52") suits them better than a 21-speed derailleur.

Variable gears

Despite the merits of single-speed transmission, for most purposes most of us prefer a choice. But how many speeds do we need? How much difference should there be between each speed? How high should the highest be, and how low the lowest?

How many speeds?

As early as 1902, when most variable gears offered a choice of only two speeds, it was technically possible to create a 42-speed transmission. Today it is entirely feasible to have an 84-speed bike without using any specially made components. Even cheap off-the-peg bikes commonly have 21-speeds. Yet some of the greatest rides have been done by cyclists using between one and four speeds. Is it a case of more speeds the better or what?

The first thing to note is that when we speak of 21 speeds we are not talking about 21 distinct choices usable in sequence. We'll look at this point in more detail when we discuss derailleurs below. What is more important than the number of gears is the difference between each and the overall range.

How widely spaced?

Common sense suggests that evenly spacing the gears is a good idea. Hence it might be thought that a spacing of 10" difference is good for touring or mountain biking. This could give a spread of 25, 35, 45, 55, 65, 75, 85 and 95". However, this ignores the fact that humans tend to perceive differences as a *proportion* of what they last experienced. Therefore it is better to work on a percentage increase for each gear shift. We would do better to suggest a maximum 20% increase, which would give a progression such as 25, 30, 36, 43, 52, 62, 75 and 90". Whilst the overall spread is almost the same as with the 10" rule of thumb, the steps would feel much more evenly spaced – despite the spacing being 5" at one end of the range and 15" at the other.

For racing purposes much closer spacing is needed to allow subtle fine-tuning of the rider's output within a narrower speed range. Increases in the order of 7 to 10% would be appropriate.

Utility riders, for whom consistency of speed is less important than the overall gearing range, typically cope quite easily with increases in the order of 33% – as provided by a Sturmey-Archer three-speed.

Incidentally a 33% increase gives a 25% decrease when changing down again. Likewise a 20% increase gives a 17% decrease, and a 10% increase gives a 9% decrease. In other words, the closer the spacing, the less proportionate difference between upshifts and downshifts.

How low and how high?

As for the range, it's horses for courses. As a starting point, single-speed utility bikes tend to have a gear around 64". British three-speed utilities traditionally offer something like 48, 64 and 85". However, this is really too high for many riders. It would be better to drop the range to something more like 41, 55, 74". This is easily done by fitting a larger rear sprocket.

Tourers benefit from a wide range, but again, it's better to get the low gears right. You can choose to freewheel down the hills but you always have to pedal up them – or get off and push. A bottom gear of about 20" is desirable for loaded touring in hilly terrain, whereas the top would typically be around 90". Its worth noting, however, that when Colin Martin successfully cycled from the UK to Australia in 1970 his highest gear was only 68". He found it quite adequate.

Road racers will typically be more concerned about the high gears than the low. Top might be something like 115" and the bottom anything from 40" upwards.

Direct and indirect variable gears

The history of cycle gears is almost as old as the bicycle itself. Since 1869 many different transmission systems have been tried. However, during most of the 20th Century two principle types have dominated all others – epicyclic hub gears and derailleurs.

With *hub gears* the rider's power is transmitted to the rear wheel indirectly. Just like a single-speed bicycle, the drive chain turns a single sprocket on the rear wheel. However, in anything other than direct drive (which is middle gear on a three-speed) the drive then passes to the hub shell via a series of meshing gear wheels in the hub. These enable the rear wheel to rotate faster or slower than the sprocket. Although this meshing introduces some friction (and therefore some loss of power) the gears are neatly enclosed in a sealed lubricated drum, protected from the weather. Hub gears therefore tend to be reliable, long lasting and low on maintenance. 'Friendly friction', as cycle journalist Geoff Apps puts it.

Hub gears, like single-speed transmissions, usually use a ½" x ⅛" chain. That is, a chain with a pitch (distance between the rivets) of ½" and a nominal clearance between the inside faces of the side plates of ⅛".

With *derailleur gears* the chain drive goes direct from a chainwheel to one of a series of different sized sprockets on the rear wheel. The rear changer mechanism coaxes the chain from one sprocket to another. Hence drive is always direct. The fact that the chain is usually running out of line reduces the efficiency somewhat, but modern derailleur chains are very flexible and the losses are relatively small.

Derailleur chains are narrower than those generally used with hub gears and are all nominally ½" x $3/32$". However, the *outer* width of the chain varies from about 8mm for use with regular six-speed freewheels to as little as 6.8mm for eight or nine-speed blocks.

Hybrid gears combine two or more different gear systems. Typically a hub gear is used in conjunction with a rear derailleur.

Let's look at the various systems in a bit more detail:

Hub gears

How hub gears work

Compact hub gears have been around since the 1890s. The first design was American but it was the Briton William Reilly's two-speed gear, The Hub, that made the commercial breakthrough. Reilly also designed the first Sturmey-Archer three-speed, launched in 1902. This led to hub gears, and Sturmey-Archer in particular, dominating the British market for decades.

Hub gears almost always work on the epicyclic or sun-and-planet principle. The commonest form has a fixed pinion (the sun) on the wheel axle. Meshing with the sun, and rolling around it in a cage, are three or four planet pinions. These in turn mesh with an internally toothed ring, or annulus.

If the drive from the rear sprocket is fed to the planet cage, the annulus will turn faster than the planet cage. This is because the planet pinions pick up additional motion from meshing with the sun and relay it to the annulus. (The size or number of planet pinions is purely a practical matter and does not affect the gear ratios in any way.) The amount of additional motion is in the ratio of the number of teeth on the sun to the number on the annulus. So if the sun has 20 teeth and the annulus 60, you get an increase of 20/60 = 33.3%. Therefore, if the annulus is connected to the hub shell, the wheel will turn at 1.333 times the speed of the rear sprocket.

Conversely, if the drive from the rear sprocket is fed to the annulus, the planet cage will revolve

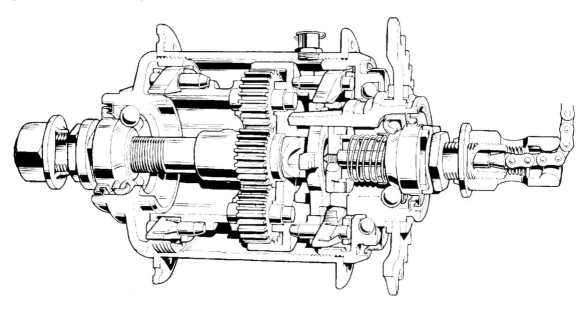

A nice way to use gears on a bicycle, the three-speed hub gear has worked well for 100 years.

more slowly. The reduction is the inverse of the increase described above. That is, 1/1.333 = 0.75 times the speed of the rear sprocket, a reduction of 25%.

Of course, it is possible to by-pass the sun and planet system and take the drive from the sprocket straight to the hub shell. This gives direct drive.

Hence a simple epicyclic gear train forms the basis of the familiar wide-ratio three-speed, typically giving an increase of 33.3%, direct drive and a decrease of 25%. Note that although the spacing of these gears looks very uneven, they are in fact *proportionately* perfectly evenly spaced, both being big upshifts of 33.3%.

The sprocket is connected to the hub shell, annulus or planet cage by a system of clutches. These may be dog clutches, which usually necessitate a 'no-drive' gap between gears. Another way is by tripping sprung pawls in and out of action, which can give seamless shifts. The familiar Sturmey-Archer AW three-speed uses a dog clutch between top and middle gears and pawl tripping between middle and low. This is why you can easily get a (possibly painful) no-drive situation between top and middle if the gear is not correctly adjusted!

Hub gear shifting can be via a handlebar-mounted trigger or twist-grip. Both ideas have been around since the earliest times, and both come and go with fashion.

One of the great boons of hub gears, particularly in traffic, is that they can be shifted when the bike is stationary.

Hub gear choices today

Today the principal hub gear makers are Sturmey-Archer (now owned by SunRace of Taiwan), Shimano in Japan and SRAM, the US-based multi-national. SRAM acquired Sachs of Germany in 1997 and are developing the Sachs range of hub gears under the new name Spectro.

Most hub gears are available combined with hub brakes of one type or another. Although wide-ratio three-speeds are still popular in some markets, there are now many other options. Shimano offer a four-speed, Sturmey and Spectro both offer fives, all three companies offer sevens, and, until recently, Spectro did a 12-speed. The small German family firm Rohloff even produce a 14-speed hub.

The five-speeds are wide-ratio and effectively like two three-speeds in the same shell. As each three-speed offers direct drive, the total spread is only five rather than six gears. Five-speeds have two sun pinions. The planets are stepped so they can mesh with both suns. To save weight and complexity, only one annulus is used.

Using the same principles, the wide-ratio Sturmey and Spectro seven-speeds are like *three* three-speeds in one shell. The Shimano, however, is a medium-ratio gear and works in a very different way. It is like two five-speed hubs in the same shell *driving each other*. That is, giving potentially 5 x 5 = 25 gears less duplicates, from which the designers have selected a smooth sequence of seven. Hence 3rd gear is 2nd x 7th and 4th is 2nd x 6th.

This is theoretically less efficient than the Spectro and Sturmey sevens. It may account for why the Shimano has no direct drive, which might highlight the extra losses in the other gears. However, the Shimano has an excellent shifting action and better spacing between gears than its rivals, although its range is not much wider than a five-speed hub.

The Spectro 12-speed had a range of 339% but was enormous, weighing three times as much as a five-speed hub. But at least they got there before Shimano, which may have been the point!

Rohloff's very expensive Speedhub is in a class of its own. It offers

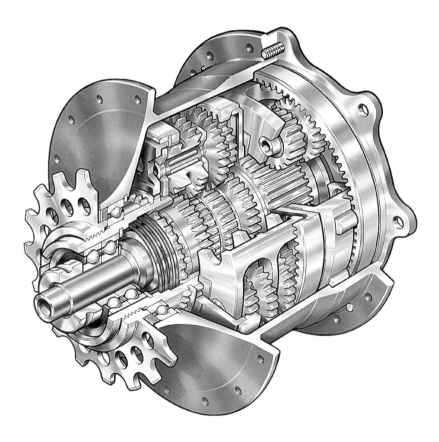

Rohloff Speedhub: Even nicer, and even more gears than a Sturmey Archer 3-speed.

14 speeds with even increments of about 13.6% and a massive range of 526%. It weighs only some 300g more than a comparable high-spec derailleur system. And its claimed efficiency of 95-98% is much better than any previous wide range hub gear. Most hub gears, other than two- or three-speeds, are much less efficient.

Hub gear efficiency

The efficiency of hub gears increases considerably the harder you work them. Ambling along on the flat at around 6mph the efficiency of the indirect gears of a three-speed can be as low as 80%. But up the pace slightly and the efficiency increases very rapidly. At fast touring or racing speeds a typical three-speed will be about 94-95% efficient in the indirect gears and 98% in direct drive.

Efficiency tends to decrease the wider the difference from direct drive. Generally the problem is greatest with gearing down. Bottom gear of a five- or seven-speed hub can be less than 90% efficient, and the extra drag can be quite noticeable.

Gear spacing and range

We've already noted that a three-speed has perfectly evenly spaced steps (all two of them!) but that the steps are very wide compared with our touring optimum. What about the five- and seven-speeds?

The Sturmey five has a range of 225% and gives relative upshifts of 18%, 27%, 27%, 18%, so the two middle shifts are pretty wide for touring purposes. The Spectro five is fairly similar.

The original Sachs seven had a range of 286% and gave 14%, 19%, 24%, 24%, 19%, 14% which is much better. The Sturmey gives almost the same shifts. But SRAM have now re-engineered the Sachs seven and the resultant Spectro product has a slightly wider range of 303%.

The Shimano seven gives 17, 14, 18, 16, 17, 16% – by far the best progression, though with a narrower range of 246% and probably lower efficiency.

Derailleurs

We are all familiar with the derailleur. Walk into a bike shop today (at least in the English-speaking world) and the chances are that 90% or more of the bikes will be derailleur-equipped.

Derailleurs have been around for about the same length of time as hub gears. Much early development work was by British inventors. However, the success of hub gears at the beginning of the 20th century led to British cycle engineers regarding derailleurs as somewhat quaint and crude devices.

The French, with more mountainous terrain to contend with, continued developing derailleurs, principally because they offered a wider range of gears. Slowly derailleurs became accepted for racing and touring purposes. However, it was not until the early 1960s that their use in these fields became universal.

By that time the number of sprockets on the rear hub was typically five which, with a double chainwheel, gave a nominal ten speeds. However, for smooth running the standard advice was (and generally remains) never to ride with a crossed chain – that is, with the chain on the larger chainwheel and biggest sprocket, or the smaller chainwheel and tiniest sprocket. Hence the range was reduced to a maximum of eight speeds, which might well include some duplicates or near duplicates. Also the difference between top and bottom gear was often not that great – many ten-speed racers had a range no wider than a three-speed hub gear.

And although indexed shifting had been tried on and off for many years, it had never worked well enough to catch on. The net effect was a complicated, difficult-to-shift system. Hence derailleurs made little impact on utility riders.

How things have changed! Although the basic workings of the derailleur are self-apparent and have not changed for a long time, it is the attention to detail and continual refinement that have made the difference. Numerous refinements, including the slant parallelogram rear changer mechanism developed by SunTour, enabled Shimano in 1985 to launch SIS – the first reliable indexed shifting system. Modified tooth profiles and little ramps on the side faces of sprockets help the chain shift more smoothly. Chains themselves are more flexible and have chamfered side plates to aid shifting.

Also, the number of sprockets on the rear wheel has crept up to 10 – with 7-9 being common. Sprocket sizes, once generally limited to a range from 13 to 28 tooth (with occasionally 11 and 12

There's eight cogs here. By the time you read this, you can already get ten...

being offered), now have outer limits of 9 and 34. And while double chainrings at the front remain standard for racing, the influence of the mountain bike has made triple chainwheels commonplace. You can even have four, if you wish.

But how to work out what all these gears are? There are, in fact, a number of configurations, each with merits and demerits. They have wonderful names – crossover, half-step, alpine, half-step plus granny, crossover plus granny, wide step, wide-step triple and rhumba. (Terry Pratchett meets Come Dancing?) Most bicycle designers/specifiers haven't a clue what this means, but if you are interested, see Frank J Berto's 1988 book Complete Guide to Upgrading Your Bike.

Gear spacings and range

The typical mountain bike today has triple chainwheels with, say, 42, 32 and 22 teeth. This set-up gives upshifts of 45% between the inner (smallest) and middle rings, and 31% between middle and outer rings. So the front changer is the one to use for big changes. It alone offers a range of around 190%.

On all but the cheapest bikes the rear sprocket cluster is more likely today to be an easy-to-change cassette type rather than the old screw-on style. Typically it will have eight sprockets from 11 to 28 teeth, thus giving about 243% range – that's a huge 463% in conjunction with the front changer.

The rear cluster is therefore the one for the subtle shifts. But how subtle will they actually be? If the sprockets are, say, 28, 25, 22, 19, 17, 15, 13 and 11 tooth, the upshifts between the gears will be 12, 14, 16, 12, 13, 15 and 18%. If they are 28, 24, 21, 18, 16, 14, 12 and 11 tooth, we get upshifts of 17, 14, 17, 13, 14, 17 and 9%. Both are quite good but you can see that the choice of a one tooth or two tooth difference between the two smallest sprockets has a big effect – in this case 18% versus 9%. In contrast the choice of a four-tooth or three-tooth difference between the two largest cogs has a smaller proportionate effect: 12% versus 17%.

In 1998 Shimano launched an 8-speed Megarange cluster with a range of 309%. The lowest upshift was a massive 31%, taking you from a 34 to a 26-tooth sprocket. The following year they produced another variant with the same overall range but more evenly spaced gears.

Tourer gearing has now more or less converged with mountain bike gearing. Even audax is beginning to move in that direction, as the 1999 Dawes catalogue showed.

Road racing derailleurs will typically have a much narrower range, and closer spacing between the gears. The chainwheels will be probably be 52 and 42 (24% upshift). The rear cluster might well have eight sprockets from 12 to 21 teeth, giving a 175% range. The total range of front and rear derailleurs combined would be about 217% (about the same as a five-speed hub). The difference (apart from in efficiency and lightness) would be in the close spacing of the gears, and therefore the wide range of subtle shifts offered.

If the sprockets used were 21, 19, 17, 16, 15, 14, 13 and 12 tooth, the upshifts would be 11, 12, 6, 7, 7, 8 and 8%. Switching the 19 tooth for an 18 would change the first two upshifts to 17 and 6%. This might be preferable, as it maintains a smoother progression, leaving one big drop at the end of the range when downshifting in hilly terrain.

Derailleur Efficiency

Disappointing though it may be to those of us who love hub gears, derailleurs certainly have the edge for efficiency.

At very low speed their efficiency in the most extreme gears (maximum chain deflection) can drop to about 88% but this is significantly better than a hub gear under the same conditions. However, as with hub gears, as the pressure is piled on, efficiency rapidly rises. At racing or fast touring speeds derailleurs can be better than 96% efficient running out of line, and about 98.5% with the chain running straight.

Hybrid gears

Hub gears combined with derailleurs

Torn between derailleurs and hub gears? Then why not have both? Hybrid gearing has been around a long time. Some people used to put two dished sprockets back-to-back on a hub gear, fit a derailleur rear mechanism and double up their gearing. Cyclo used to produce two and three-sprocket converters to fit hub gears. Dave Connley's Dacon converters allow four derailleur sprockets to be fitted to a hub gear. Used with a five-speed hub, these nominally give 20 speeds.

Another approach is to use a double or triple chainwheel with a hub gear. Chain tension must be taken up, either by a standard derailleur rear mechanism, or by a special tension arm.

In the 1980s Sachs marketed a three-speed hub complete with two sprockets for use with a derailleur. They also launched the Orbit miniature two-speed hub which took six sprockets, thus offering a 12-speed transmission without the need for a front changer.

Today SRAM market the Spectro 3 x 7 system, whereby a three-speed hub is fitted with seven derailleur sprockets. The system gives a very wide range of gears, plus the ability via the hub gear to make large shifts with the bike stationary.

Hybrid gearing combines the advantages, disadvantages and inefficiencies of derailleurs and hub gears. Whilst some riders will find it very satisfactory, for others the complexity and weight will be off-putting.

Mountain Drive

The epicyclic system used in hub gears can also be used in the bottom bracket and/or chainwheel. In the early 20th century bottom bracket gears were favoured by some cycle manufacturers, notably Sunbeam, James and Centaur.

The Schlumpf Mountain Drive is a modern Swiss two-speed epicyclic chainwheel gear. It comes as a complete chainset and bottom bracket axle/bearing replacement. Type I gives a huge downshift of 60%, whereas Type II provides a 65% upshift.

The Mountain Drive is expensive. However, used in conjunction with a hub gear and chaincase, it gives a very wide range with the transmission fully enclosed. Development continues and the 1999 version can be shifted under load.

Electronic shifters

Recently electronic shifting systems have become available both for derailleurs and hub gears. Mavic introduced their Zap system for derailleurs in 1995. Instead of mechanical cable linkages, it used an electric shifting system. This meant the rider could have duplicate shifter switches – good for tri-bars. Wet weather, however, adversely affect the Zap's wiring, so after another three years of development, Mavic launched the Mektronic. Like your TV remote control, this uses infrared beams to link the shifter switch to the derailleur mech.

Systems like the Mektronic lend themselves to automatic microprocessor control if that's what you want. So shifting could be linked to wheel speed, pedal cadence, heartbeat or some relationship between these factors. Shimano's Nexus Auto D system makes a modest step in this direction. It automatically shifts a four-speed hub gear according to wheel speed. The rider can opt for sports or cruising modes and can also manually select gears.

Conclusion

What gear you choose will be dictated by cost, proposed use, availability, fashion and your temperament.

For commuting and utility riding, hub gears have much to commend them. Although very out of fashion, the humble three-speed offers a useful range, relatively high efficiency, low cost and reliability with minimum maintenance.

Five and seven-speed hubs offer a wider range and more comfortable spacing between gears without adding too much additional weight. Efficiency, however, tails off badly in bottom gear. Some riders nonetheless find them suitable for touring, especially if geared low. The 14-speed Rohloff Speedhub seems to overcome the limitations of other hub gears, although its price will deter all but the wealthiest connoisseurs.

For most other uses the derailleur configurations outlined above are hard to beat, offering the best balance of efficiency, overall range, spacing between gears, weight and cost.

And don't forget: there may just be a case here and there for single-speed!

The Mountain Drive two-speed bottom bracket. A little specialised, but some get on well with it.

Suspension – the Good, the Bad and the Bouncy

It takes at least a year to become a suspension expert – and another two to realise that you don't have a clue.

A quick bit of history. Suspension development can be divided into three distinct phases. In the 1880s, many of the early 'safeties' used suspension to compensate for their 'small' wheels. This was undoubtedly the most successful period, as it resulted in the pneumatic tyre: still the lightest, cheapest and most efficient form of suspension. Second was the Moulton, the first and probably still the only bike designed by someone who actually understands suspension. The third phase is the one that I and the rest of the world are obsessed with – mountain bike suspension: to date the most expensive, energetic, over-hyped but least successful of the three.

Why suspension?

To understand why we are not having much success, you need to understand the reasons for adding a lot of extra weight and complication to the normally uncluttered bicycle.

Foremost is the need for comfort, for we are not talking here about gliding along on the old black stuff, but about crashing over the bumps and ruts of our wrinkly old planet's natural, or near-natural surface. Comfort is not just about an enjoyable ride, it is even more about reducing rider fatigue – just as important to leisure cyclists as to racers.

Second is control – keeping a bicycle pointing in the right direction at 50mph (80kph) down a rocky mountainside is not easy, and it gets even harder if the wheels are not touching the ground!

The third advantage is speed over the ground. Obviously, the previous two considerations have an effect on this, but if wheels are able to stay in touch with the ground rather than jump over obstacles, the ride is more energy efficient, for any energy expended riding 'up a bump' is the same as riding up a hill, but you get it back only if you roll down the other side. Unfortunately, any damped suspension system will, by its nature, absorb energy: shock absorbers will get hot and you don't need three guesses to work out where the energy has come from.

So that's why you need the bounce – the trick is to do it without adding too much to the weight or cost, without reducing the lateral stiffness, or having the multitude of extra bearings wear out in a few months. Most of all a way has to be found to arrange for the wheels to go up and down over the bumps, but not move at all when you push on the pedals!

Types of rear suspension

To discover how successful we have been to date you only have to visit an off-road event, where you will see an abundance of suspension forks and very few other forms of suspension. To really see what is on offer you will have to visit one of the big shows, like Cologne or Anaheim, where you will see an apparently infinite variety of solutions to the problem. Having studied all of these for some time now, I have realised that most rear suspension systems fall into one of four main categories:

Simple pivot designs
These fall into two sub-categories: high-pivot types, usually using massive swing-arms, and low-pivot types, usually having a traditional rear triangle.
Pros: Simple.
Cons: High pivot is affected by chain tension, can be heavy. Low pivot design relies on suspension unit for lateral location.

Four-bar linkage mechanism
A four-bar, or if you're clever, a three-bar linkage mechanism, is similar to the above but it always has a lowish pivot and relies on a pivoted plate to link the suspension unit to the upper stays. This

Simple pivot designs

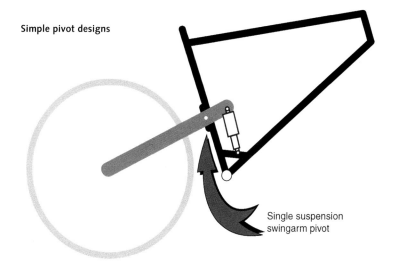

Single suspension swingarm pivot

allows the unit to be situated almost anywhere, hence the seemingly endless variety.

Virtually all of these designs have a pivot near the rear dropout, either in the upper or lower stay, which is not actually necessary if you arrange the set-up correctly, but very few seem to realise this. Notable exceptions are DPS with their F1 design and me with my current design (which is a copy of the F1!).

Four-bar linkage mechanism

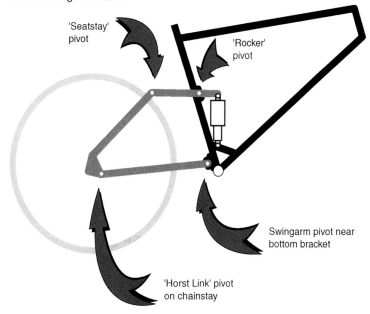

'Seatstay' pivot

'Rocker' pivot

Swingarm pivot near bottom bracket

'Horst Link' pivot on chainstay

Pros: Suspension unit can be positioned anywhere.
Reasonable lateral stiffness.
Can be arranged to give a 'rising rate'. In other words, it gets progressively harder.
Cons: Lots of bits.

Unified rear triangle

This is where the bottom bracket is part of the rear rather than front end, which totally eliminates the chain tension problems but, of course, gives you some new ones – the saddle-to-pedal distance can now vary, and when standing on the pedals you lose a lot of suspension. Also, the pivot bearing loads are increased.

Pros: No chain tension problems.
Cons: High bearing wear.
Complex/heavy.
Saddle-to-pedal length change

Unified rear triangle

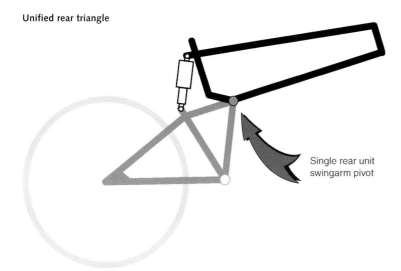

Single rear unit swingarm pivot

Parallelograms

The only true four-bar linkage systems, typified by the Mert Lawill design with two pivots above and below the chain line at the front. This virtually eliminates the tension problems but allows for a very responsive system. One new variation is the Rock Shox/Tomac bike that has a very compact parallelogram mounted just above the bottom bracket and has a 'unified' rear. This is still experimental and looks simple, but I suspect is not.

Pros: Good, active suspension.
Cons: Lots of bits, high bearing loads.

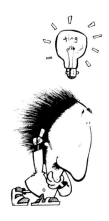

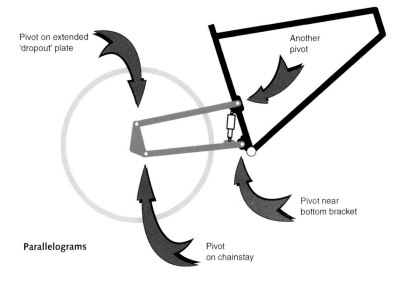

Parallelograms

▶ eating the bounce

All of the above systems cope with the perceived problems of off-road riding to a greater or lesser degree. What some of them cope less well with, and what nobody ever talks about is 'rider-induced bounce' caused by the vertical motion of the rider's body, and weight transfer between bars and pedals when honking. The one way I know to deal with it is to turn the suspension off, which brings me to the suspension medium itself.

Rubber elastomers are simple, cheap, light, and long-lasting. They can be quite effective – examples are the Proflex bikes and of course Dr Moulton's very clever little rubber ball, which is still far better than very many more complex systems.

Secondly, the current flavour of the month – the coil spring with oil damping – is heavy, expensive, and can give a real Rolls-Royce ride. It allows for a good deal of 'tuning' and can be designed to have a lock-out function.

Then there is air and oil : technically the best, but out of fashion due to 'stiction'[11] and overheating, although this is probably more

---

[11] *Stiction = 'sticking' due to friction, meaning that it takes a biggish bump to get the suspension moving.*

imagined than real. Units are lighter than spring, more expensive, and give even more control options – even the rate of suspension rise can be pre-set, and, by controlling oil flow, damping can be varied.

Front suspension

All of this so far relates to rear suspension. The front end is in fact rather more important from the point of view of function, but is dominated, as is the motorcycle industry, by the technically suspect but actually very effective sliding pillar fork. These also come in the same options of elastomer, air, spring, oil in many combinations. Most perform remarkably well. Some work better on fine ripples, others are better at absorbing the big bumps. Interesting, but not as yet very popular, are Cannondale's headtube system, and the AMP and Girvin leading-link forks, which are stiffer but do not offer the long travel options of the telescopics.

Mystic Mike

What about the future? Whilst current trends suggest that soon all off-road bikes will soon have full suspension, the more of them I make and ride the less convinced I become!

Downhill from here

One of the problems with writing a book, as opposed to the magazine articles I have been scribbling up until now, is trying to make it timeless. This should not be a problem for chapters like the one on aerodynamics, based as it is on well-established data and only a *few* wild guesses. But for suspension, where the wild guess is the equivalent of a wind tunnel, it is impossible. To give you some idea of this, I have left the previous part of this chapter as it was written for 'Bike Culture Quarterly' magazine in 1996. Writing this now in September 1997 I realise how hopelessly out of date it is (or even was). For apart from the inevitable small technical changes, there is no mention of 'downhilling'.

This should tell the reader a lot about my ability to identify a trend. I did not spot this one until the spring of '97, by which time downhill was the biggest thing in cycle racing in the UK, USA and several other significant bits of the planet.

And along with downhill has come its unruly offspring 'free-riding' – a marketing person's idea of a label for people who play about in the woods on their bicycles. The label is a bit sad really,

Sliding pillar fork: bad in theory, good in practice.

This is the one you have to push uphill. They're called downhill mountain bikes for a reason.

but the idea is great, as playing in the woods is just what dual suspension bikes do best.

Nobody is obsessed by weight any more. Typical downhill racers weigh 18kg, so it is a lot easier to engineer some decent suspension pivots, etc. Most of the stuff about 'bio-pacing',[12] active suspension[13] and the like is forgotten, as gravity is now no longer the enemy but the main source of power. And, of course, suspension travel has shot up – 150mm front and rear is now the norm in competition.

One thing that seems to have improved is the quality of the damping/suspension units and their tuning and adjustment. Yet front forks are in many cases larger than on many motorcycles, more for image than necessity. Also for image is the almost universal use of the triple clamp fork crown system. This is not only unnecessarily heavy but its use means that the bike's weak link is unlikely to be the cheap and easy to replace steering column but will in fact be the not at all cheap frame. This is obviously stupid, but it is the fashion, so what can you do?

---

[12] *Rider-induced suspension bounce. The term derives from Shimano's Bio-pace asymmetric chainwheels of the 1980s.*

[13] *This is suspension that remains 'active' while the rider is pedalling.*

The most popular suspension medium currently is elastomers for the forks and coil/oil for the rear end. Top end forks use coil/oil, and Rock Shox are about to reintroduce air/oil for cross-country forks.

Smart damping

A very interesting new development is the idea of 'smart' damping, now available from Proflex. Several ways have been suggested but most likely to succeed is electro-reological fluid. This is a liquid that changes its viscosity when an electric current is passed through it – and it does it in three milliseconds! This fluid in the shock unit could be linked to a series of sensors and even a microprocessor. It would be able to adjust the damping as required while you rode. Anything from zero damping to full lock-out – the possibilities seem endless. The down-side will be not just having to wash the bike at the end of an event but download information from it as well!

This could be the way to make dual suspension work for cross-country. Nothing else seems to have done. The winning is still achieved with 'hardtails'. The suspension forks are really only necessary to keep the front wheel on the island on the faster descents. (Although I would probably not be nursing a broken collar bone if I had used a pair last week. Ho hum…)

Spring in the city?

Away from the mud and mountains one unfortunate trend is the appearance of large wheel (700C) city bikes with suspension, often with aggressive non-diamond styling. This has come about as a result of marketing getting the upper hand. It has absolutely nothing to do with us engineers, such machines being heavy, expensive and unnecessarily complicated – and not even any more comfortable. For as you will have noticed, it is the nature of cities to be provided with roads rather than trails. And while we might complain about their condition, the occasional pot-hole does not demand dual suspension. The 'discomfort' that these bikes are intended to cure is largely caused by the rider's weight on the saddle and the best cure for that is a good saddle and plenty of exposure to it – or a recumbent.

And now...

At the end of the earlier section I attempted to predict the future of suspension. I will not repeat that mistake, but I will make some

Not my idea of a good bicycle.

suggestions as to where I think suspension should go.

Firstly, for city use small (20") wheels and suspension do work very well together as Dr Moulton has demonstrated. The problem is, most of us cannot afford his beautiful bikes[14], so I guess I shall have to talk Giant into making one. As for the big and bouncy off-roaders, where they are going seems fine. They deliver fun in large doses, allowing you to do things on a bike that you would not normally think possible – or even desirable! However, like the recumbent, they are not a replacement for your regular bike but an excellent addition to any 'collection'.

This is where you need the bounce – off road. It's now essential even for cross country riders, and has been since Paul Lazenby won on it.

And there's more

It turns out that getting a book published takes even longer than writing it. So here's a timely bit of late news, as of late 1999.

First change is that downhill or 'free-riding' is not so much dying as being stabbed in the back by the majority of cycle journalists. They seem slightly embarrassed that they ever had anything to do with it. This is a shame, as I am sure that most people who bought downhill-style toys knew what they were getting, and had no illusions about their usefulness as transport. The new bandwagon is baby (14") hardtails with 4" travel forks – sort of dual slalom/trials machines, great for street play. What's left of free-ride is now a lot closer to cross-country than downhill.

Which brings me nicely to the best bit: straightforward cross-country dual suspension – light as can be and meant for racing. Everybody's doing it. It was started in the UK by Marin, whose works rider Paul Lazenby won the national points series on a production 'boing' bike. Specialized were not far behind with the new super light FSR series, also Scott and now Trek and Fisher. GT, on the other hand, produced the 'I drive', which is the emperor's new clothes in cycle form.

I am pleased to say that my own company meanwhile have made a bit of history, even if it was without my help. Instead they had the assistance of a good-looking young French mountain-biker, Pascal Tribotte. His day job just happened to be designing bits of Formula 1 cars for Renault Sport. Getting bored with the 230 mph supercars, as you do, Pascal started doodling rear suspension for mountain bikes. This must have been more challenging, as F1 cars only have 12mm of suspension travel. Anyway, Pascal doodled well and the result is the XTC, a truly smart bike but with not a microchip in sight.

---

[14] *Dr Moulton's 20" wheel machines, licence-built by Pashley, are the least expensive of his designs in current production.*

Giant's full-suspension cross country bike. 'Totally brilliant, but I didn't design it. Doh!'

What Pascal realised was that: whilst the force that caused the rear wheel to compress the shock absorber – a bump – was much the same as the force created by the weight transfer and inertia of feet and legs pushing on the pedals, the force that then resulted from the foot's action (the chain tension) *operated at right angles* to the other forces.

Not that Pascal was the first to realise this, of course. Many high pivot designs rely on this chain tension to hold down the rear end. But such a system is seldom in balance and either bobs or makes the suspension a lot harder than desirable. Pascal's solution was to run the system 'topped out', that is with the rear shock at a pressure that prevents any sag when the rider sits on the bike. This means that the chain tension is trying to push the suspension in the same direction that the shock has already done, so no further movement is possible. But it still does not lock the system out – it just balances the forces. So any bump hit, even when pedalling hard, will be absorbed. (See figures for more details.) Better yet, get yourself a test ride – it really does work. Sorry to sound like a salesman but it is something special, even if I did not design it!

Getting to grips with operation of the XTC system finally gave me an understanding of a system that's been used by both Specialized

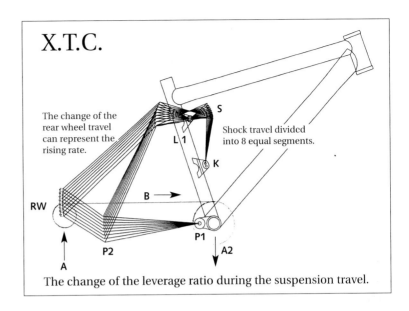

The change of the leverage ratio during the suspension travel.

Right, are you sitting comfortably? Then I'll begin. Imagine you are riding along on your revolutionary XTC. You ride over a small bump. This will produce a force shown as A, that will cause the suspension to compress the main point of rotation, P1. Likewise, if you push on the pedals the inertia forces will push down on the frame and a similar rotation about P1 will occur. Except that, at the same time, the force that you are creating is turned at right angles by the chain, resulting in force B. This will have the effect of rotating the seatstays forward around point P2. This applies a force to L1 from above rather than below, due to its angle of rest (top-out position). This balances almost perfectly the force A2, thus creating a true Force Balancing geometry, which is my description and so far my only input to the project. Not that I am in any way bitter...

and GT. This is the Horst Leitner link. It looks a lot like the XTC but is actually not the same at all. It gives a similar modified wheel path, which reduces the chain tension input, thus reducing bob – but not it does not eliminate it.

Other trends are towards air springing for both front and rear shocks, and using air pressure both sides of the shock absorber piston (air negative spring) to make things smoother. I *think* I understand. All of which helps makes the new wave of cross-country bikes even better.

When I wrote about suspension earlier, I was rather sceptical about it ever replacing the hardtail for performance riding. Not any more. As far as I'm concerned, it has replaced the hardtail for off-road use. Dual suspension is all I ride now and it is all you will want to ride.

11. Retarding Progress

Retarding progress: no, not the UCI[15] but brakes, which are almost as bad, as they don't actually make you go any faster. I wasn't going to mention them, but my co-conspirators twisted my arm, so here is a reluctant review of how *not* to go fast.

Velodrome

Simplest by far is a 'fixed wheel'. It's all you are allowed on a velodrome, and it's also ideal for time trialling on fast courses. There's not only no rear brake, brake lever or cable, but no rear mech, lever and cable – so it's cheap, light, aerodynamic and efficient. On the road, both the law and common sense require a brake on the front wheel. My own favourite is the Campagnolo Avanti, which is a very neat and cheap little side-pull that would go well on both ends for those who cannot do without a freewheel and gears.

Road racing

For road racing, where brakes do make a difference, the only choice is the current Campag or Shimano. They both use the same double-pivot design, and are both very well made. They also function superbly, especially in the Record or Dura-Ace versions. To operate them (and change gear) the matching Ergopower or STI levers are equally a must for road racing. The extra weight is a small price to pay for such a perfect function.

Time trial, triathlon and touring

For time trial or triathlon combined levers are not so useful, and you can achieve better results with separate gear and brake levers. They

Sidepull brakes: fine for road bikes

---

[15] Union Cycliste International, traditional cycle racing's ultra-conservative governing body.

are lighter and cheaper, and the cable runs are usually easier to arrange.

For touring the traditional cantilever is still the best bet. Though not as powerful as the new V-style brake, it is a lot more progressive and tolerant of wear.

The only way to stop off road – the disc brake.

Off-road

V-style brakes do come into their own for serious off-road use. Though their extra power can be very useful and the lower lever pressure is a blessing on long descents, they are fiddly to set up adequately and wear out rims even faster than cantis, especially in muddy (British?) conditions. They also need a good true rim with no dents, and they can be a bit abrupt.

To solve all of these problems what you need is disc brakes. Not a new idea but about to burst onto the mountain bike scene in a big way. I was a late convert to discs, thinking it was all a bit unnecessary and expensive. That was before I used them. Now I would put them on my butcher's bike – if only Hope did a mounting kit for a 1950 Gundle Low CG!

Discs are a little heavier than Vs and at the moment are a lot more expensive. But they give very good stopping power in all weather conditions, are wonderfully progressive, never clog, don't need setting up when you change pads and are unaffected by dented or buckled rims. They are already a must for downhill racing, and the Giant team are now using them for cross-country. Others are set to follow. To date I have only used those made by Hope but there are many other excellent ones coming on the market – Magura and Formula both look good.

Fitting them yourself can be a problem. Most suspension forks are now appearing with mounting points, but fitting the rear calliper is tricky on some frames. More frames need to include mounting brackets.

Between writing and publishing this book, discs arrived in force. They now come in all shapes and sizes from Shimano to obscure. Most work well, though some cable operated ones seem a bit dodgy, so look out for road tests in magazines.

The unbeatable drum?

My own favourite 'stopper' is the drum brake. There was a time when I went around like a prophet proclaiming the drum as the answer to all our problems – from time trial to mountain bike. But I was wrong. They are not powerful enough for off-road, and under extreme conditions will let in water. They then provide about as much stopping power as a badly adjusted dynamo. They are also not much use for time trialling. They stop well enough and look really neat on one side of a monoblade. But aerodynamically their sheer bulk is bad news. Far better an untidy looking calliper brake, especially as it will be mounted on the front of an equally 'messy' fork crown, thus adding little to the overall drag.

Recumbents

Drums do work well on recumbents, especially trikes. The smaller wheel diameter means more stopping power. Rim brakes are virtually the same for any diameter of wheel. However, any brake operating on the hub is fighting against the leverage of the wheel's radius, so large wheels are hard to stop and vice versa. Also, as you don't need the rims for stopping you can 'disc' the wheel with clip-on or glue-on fabric – and that really is aero.

City use

But what drums do best is go shopping or commuting, where their smooth progressive power is just what is needed, and where maintenance is more like every ten years rather than every ten rides. Best of all is to use them in cantilever mode on monoblades front and rear. They then protect both the brake and bearings from the elements, as the outer bearing can be totally sealed. And as punctures can be fixed without removing the wheels, they are seldom disturbed.

Alternatives to the basic drum for city use are the back-pedal or coaster brake, still popular in Germany but little seen elsewhere. An updated version, the roller cam, is now being produced by Shimano. It is available for front or rear, and on the rear is available as hand or back-pedal operated.

The drum brake: the most civilised way of stopping. Ideal for city bikes.

12. Monoblades and Cantilever Wheels – the Single-minded Approach

It is not that I was the first to use them – far from it. And I am sure I will not be the last. But it could be that I am the one who understood them.

For to most cyclists the 'monoblade' and the cantilever wheel are an anathema. This, I suppose, is only natural, growing up in a world of forks and stays, with the wheel always perfectly central. And woe betide any frame-builder whose wheels did not line up perfectly. But although as cyclists we are used to seeing the wheels supported on both sides, we are of course surrounded by a world where most of the wheels are cantilever – that is, supported on one side. Indeed, there are very few vehicles other than our bicycles, their motorised cousins and wheelbarrows that still have forks. And even some of the motorised machines are abandoning forks or stays in favour of the 'monoblade', 'leg', 'stay thing' or whatever you care to call it.

Historical precedent

Where it all started: the Invincible, produced in 1889.

Most famously, the Vespa scooters of the '60s were cantilever front and rear, and carried a spare wheel under one of the side covers. Historically it is a much older idea. I first saw the idea on a bicycle when visiting the Museum of British Road Transport in Coventry in 1985, in the form of the 'Invincible' produced by the Surrey Machinists' Company in 1889. This was, in effect, a cross-frame design but with one side left off – not actually that good an idea, as what is left is not really stiff enough to support the off-centre loads. But even this machine had borrowed the idea from an earlier Coventry-built velocipede!

The idea did not catch on, and the first commercially successful use of the principle on a two-wheeler was the scooter. Since then it has appeared on the occasional 'designer' bike and on an interesting

One leg can be really cool,
as Vespas prove.

The bottom end of a Euro fighter.
It doesn't use forks either.

French ATB called, I think, the Laiti; this saw limited production. Currently several high performance motor-bikes have cantilever rear wheels, and many of the current crop of scooters have 'mono' front and rear.

Constructional convenience

I first consciously used the idea on the Mk2 'Speedy', as a way of simplifying the frame design. By mounting the single rear wheel on one side I was able to run the main 2" diameter aluminium tube right through from the bottom bracket to the rear axle, and not have to graft on a rear triangle. It looked very strange at first but worked extremely well. It was some while before I realised that it was really no different from the two front wheels, except that they were not on the centre-line, so they looked normal. For that matter, the rear wheel was no longer on the centre-line either, for it had to go on one side of the main tube. Bending that would have been tricky, so I decided that maybe a bit of asymmetry would not be a big thing. In practice this turns out to be true. The rider's biggest problem is remembering the offset when avoiding pot-holes.

My first use of the mono, on the Windcheetah trike. It's still used.

Improved aerodynamics

Despite having made this small mental leap in 1982, it was not until I saw the 'Invincible' in '85 that I began to understand the possibilities for two wheelers. I had by then developed the original monocoque racer, which was what I was riding as part of the Rover Centenary ride when we stopped off at the museum in Coventry. On the way home on the train the pieces dropped into place. This time I did not need the cantilever for *convenience* but for its *aerodynamics*. For as I explained in the chapter on that subject, bi-planes are bad and mono-planes are good. So one big fat-but-aerofoil section blade would be a lot better that a pair of regular, or even aero-section, blades.

I was still not too familiar with composite processes at this stage, so this first blade was machined and filed from a length of 2" square 6061 aluminium. This, I have to say, is not a good way to earn a living! It worked well enough, though, and I did some good rides on the bike. But it was another four years and a lot more 'triggers' to finally hang the rear wheel on one side as well. Again, it was out of line but by only 16mm this time. It was this Mk2 that was adapted by Lotus for Chris Boardman to use so successfully in the '92 Olympics.

Further development

Having got this idea of cantilever wheels and monoblades as an option stuck firmly in my mind, I built a series of bikes of different types in an attempt to discover the advantages and disadvantages of the system. I had already built a short-wheelbase recumbent bike as part of the Mk2 development programme, again using the same 2" alloy tube as on the Speedy. This again had considerable wheel offset. It worked well, and was eventually sold on to a friend who successfully raced it in HPV races.

I then built a shopper in answer to a request from Jim McGurn who, at the time, was editor of 'New Cyclist' magazine. He actually only wanted me to write an article on city bikes, but I was not in writing mood. And so 'Amsterdam' was built, brazed from largish steel tubes and with the most offset rear wheel yet – 60mm. This you could notice when you first rode it, but after 10 minutes or so it was just like a bicycle. This was followed by 'Vienna', a carbon fibre/aluminium tube rear suspension touring bike. This was more 'interesting' than successful.

Then came 'San Andreas' – well, what else can you call a mountain bike with offset wheels? It was similar to Vienna but rather better

Amsterdam: the first mono shopper.

built. Not exactly a great bike, but a lot of fun to ride, it was also sold on to a friend who has since used it three times to complete the Polaris Challenge.

I then rather more purposefully designed the 'Ratcatcher', a long-wheelbase touring recumbent. It used the same constructional techniques as the 'Speedy'. I built two prototypes, then five pre-production models that were sold to friends whom I hoped would not take me to court if the wheels fell off. The intention was to give them a year in the field and then, all being well, put them into production alongside the 'Speedy'. But before that could happen, I took the job with Giant. So 'Ratty' is on hold.

The last mono shopper?
One day it may be in production.

A Giant step...

Having got the job, and now that I am an 'official' designer, I don't do quite so many funny bikes. But I have done a couple of 'mono' designs.

From Amsterdam to Cambridge

'Amsterdam' evolved into 'Cambridge'. This was built from aluminium tube and castings, and had the multi-sprocket transmission completely encased in the frame. This was built to sell the idea to Giant of a purpose-built city bike. It has now reached 'Paris', which is moulded in one piece from glass-reinforced plastic, and has what I describe as an 'internal direct transmission system'. Currently it is being used for a comprehensive market survey. Assuming that everyone likes it, and that we can find a cost-effective way of moulding the frame, it should be in production one day.

Mighty, meaty, big and bouncy.

Mountain bikes

I also played around with ATBs, adding a glass-fibre monoblade (complete with disc brake) to the front of my hardtail; it was very light and very strong. I also built a single leg suspension version with 75mm of travel and employing the wishbone link principle, as used on virtually all the world's aeroplanes – monoblades every one! Amazingly there are people in Giant who think it might be worth trying to sell the idea – Long John Silver lives!

Conclusions

So what have I learned after 15 years of 'monoculture'? 'Swings and roundabouts' is the answer. It is a matter of understanding the reasons and advantages, but also the trade-offs.

Racing

First, and easiest to understand, are the aerodynamic reasons. If air resistance is your big problem, either on the velodrome or in a time trial, then a monoblade in place of a fork is almost a must. It will add weight, and the fact that it tends to flex sideways, rather than fore and aft, does not encourage cornering at the limit. But it will impress the man with the stopwatch.

It also makes sense to have a cantilever rear wheel in these situations. This is not an add-on option, though. But for those intent on designing the best, it's the *only* way to go. You can even fit gears – as on the 'Speedy'.

For road racing, however, it makes no sense at all. You are mostly in the bunch, so get little aero advantage – just poor handling, extra weight and a front wheel that you can't change easily if you puncture. For time trialling, of course, you have no outside help and you have to change the tube, which is easier with a cantilever wheel. The cantilever can also make sense for structural reasons, as on the rear end of the 'Speedy' and 'Ratcatcher'. Here the offset rear wheel is a small price to pay for a very strong simple frame and a much easier run for the inconveniently long chain, as there are no stays getting in the way.

Mountain biking

Also, for structural reasons of a different sort, it makes sense on the front end of a mountain bike, where it is stronger and lighter than a conventional fork. This might sound strange, as I have already dismissed the idea for road bikes. But we do not do to road bikes what we do to mountain bikes. Quite simply, we abuse them. We ride them into tree stumps (collar bone!), down boulder-strewn mountain sides, not to mention the steps outside the local C&A – and we expect them to survive!

Now it is in the nature of things that, as they get bigger, they get stiffer and stronger, and that they do this faster than they get heavier. This is all explained very accurately by Sharp. Suffice to say that, as you double the diameter of a tube, its weight doubles but its stiffness will increase some four and a half times – really nice numbers. This means that if you were to swap your two 25mm diameter fork legs for one 50mm diameter leg, you would have doubled the critical fore and aft strength without any increase in weight. You will also have decreased lateral stiffness, but that is already many times higher than necessary.

For suspension legs there can be even more advantages, as there is only one set of parts. And one big set is cheaper, and should work better, than two small ones. You will have to use disc brakes, as there is nowhere for the 'other' canti to go. But that is no bad thing.

Despite having thought long and hard about it, and even built one, I can see no good reason for a mono rear on an ATB – and quite a few reasons not to. But I have not given up completely.

City use

The last and probably best reason for going 'mono' is for the sheer convenience of it. Firstly, tyre changing for punctures. Not for racers – they have quick-release wheels – but for commuter/city use, where a quick release is an invitation for someone to walk off with your wheels, and where you may have to deal with a chainguard, hub gear and drum brake before you can get your wheel out. With a single-sided shopper you don't take anything apart. The wheel stays where it is. The drum brake need never be disturbed. And the chain and gears are inside the bike, running in oil rather than wiping it off on your clothes.

It is also a more convenient design for manufacturing. Not at the moment, as we are all a bit stuck on the diamond frame. But starting from scratch there will be fewer parts to manufacture and assemble.

Reactions

The other thing I have learned is what people's reactions are likely to be. Turn up at a mountain bike race and they will stare incredulously. Amazed that it goes in a straight line, they will politely refuse to try it themselves. On the other hand, shoppers in a town centre are unlikely even to notice that it is standing on one leg.

Having got the idea well and truly stuck in my mind, I now treat it as any other option in a totally rational way. Yet no one else seems to be following me down this particular path. *Is this my single-mindedness, or their lack of it?*

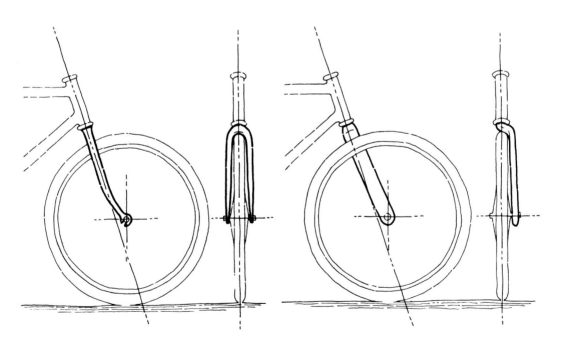

One big fat one really does do a good job. The wheel is never aware that it is mounted on one side and the stresses in the frame are the same as for forks.

13. Lubrication

Keeping the main bearing elements of the pedal cycle lubricated should not be much of a problem. They are, of course, all rolling element bearings and usually of more than adequate size. For example, the most popular ball race for top end hubs is the 12mm bore x 28mm overall diameter 6001 bearing. This has a dynamic load carrying capacity of 500kg at a maximum of 28,000rpm – figures that the average cyclist is unlikely to exceed!

An environmental problem

So if you were to ask a bearing manufacturer about lubricating a bearing operating at one tenth of its load capacity and one sixtieth of its speed limit, he would tell you merely to put a small dab of grease on once in a while. If, however, you told him the bearings were on a bicycle, he would probably deny all responsibility for his products, due to the special circumstances.

This is because the problems that cycle bearings face are not so much a lack of lubrication as an excess of 'environment'. Water causes the polished surfaces to corrode. And dirt causes abrasive wear and fatigue of the hardened steel surface, as it creates small points of ultra-high pressure.

Possible solutions

Heavy-duty seals are available to solve these problems, but fitting them would add to the cost and weight, and would increase the bearing drag. So the people who would benefit from them won't pay for them – or are not given the choice. And the racing cyclists who have the money don't want the extra weight or drag.

There are some excellent replacement hubs and bottom brackets available that allow you to pump new grease in at the centre which pushes out the old grease and dirt through the seal, where it can be wiped off. There is an element of overkill in this approach, but it will guarantee a long bearing life. Bearing drag will, of course, be higher with all that grease being churned around. In fact, for normal bearing use it would be very bad practice. But that is because of the much higher speeds – you are unlikely to overheat the grease at 400rpm.

For those worried about the excess drag of grease you could use the same idea but use a small amount of thickish (engine) oil instead. This will soon find its way to the bearings, unlike grease that does not flow at room temperature. It will, of course, also flow out through the seal eventually, taking some dirt with it and making your bike look a bit messy. Worse, if the oil gets on wheel rims it will adversely affect braking, and if it gets on tyres it will do the rubber no favours. So you will need to keep a cloth handy to wipe the bike down occasionally.

The chain problem

But all these problems are as nothing in comparison to the cyclist's greatest tribological challenge, the chain – famous for being 'rusty-dirty', 'rusty and dirty', 'oily and dirty', etc. And these problems of lubrication of an otherwise excellent device have caused more 'designers' to 'improve' the cycle by giving it a toothed belt, shaft drive or whatever. The real solution to the problem is an oil-bath chaincase. But Sunbeams have been out of production for 30 years, and my updated version will not be available until 2001 (with any luck). In the meantime you have two choices.

The proper method
To do it properly you need to:
- remove the chain,
- clean it in soluble degreaser,
- heat it with a hot air gun till too hot to touch,
- drop it into a container of thick oil (I use SAE 150) and leave for a few minutes,
- hang it for an hour or so to drain off surplus oil,
- lay it on a soft tissue to remove any remaining surface oil,
- and finally, refit it.

Then repeat as necessary!

The quick spray approach
The above method works very well but is obviously a lot of effort. The popular alternative of using an aerosol of either wet or 'dry' lubricant is a poor substitute. This is because any oil added from the outside to a dirty chain will take some of the dirt with it into the bearings, increasing chain wear and in turn chain ring and sprocket wear.

The clip-on chain cleaner

Between the 'proper' method and the quick spray approach is what may at first sight seems to be a sensible compromise. It is possible only for derailleur-geared cycles and involves using a proprietary chain cleaner device. This clips around the chain in situ and feeds it through various brushes and a small bath of degreaser. Rotating the cranks backwards about 30 times ensures that the chain is fairly well cleaned. And much of the muck and old lubricant is caught in the cleaner for safe disposal. However, almost inevitably some of the dirty degreaser drips off the chain or is sprayed about by the sprockets. This makes the process fairly messy.

But here's the real rub. Having worked degreaser into the inner recesses of the chain you then have to relubricate it, either with an oil can or an aerosol. Hence deep within the links you are mixing lubricant with degreaser, unless you dry the chain as in the 'proper' method! Use of a chain cleaner can therefore merely be a compromise solution – a not particularly quick but nonetheless fairly dirty way of achieving a second-rate result. Not surprisingly, many chain cleaners end up being used only a few times before enthusiasm wanes.

Cables

That other bane of the cyclist's life, the brake and gear cables, have thankfully become much less of a problem thanks to nylon linings and PTFE coatings. However, many cables will still rust if left unattended for long periods. The only real problem though is the last bend of the rear mech cable. The bends cause the friction and as this one is close to the muck, a lot of it gets sprayed onto the open cable above and works its way inside the cable sheath.

One solution is to run full outer cable up to the top tube. Or you could try the new caterpillar-like gizmo that SRAM have put on the market. This could be pre-packed with grease, which hopefully will work its way into the outer cable faster than the dirt from the lower end.

Velodrome minimalism

If the bike is only to be used on the rather less environmentally-challenging velodrome, you could strip down all the bearings, clean out the grease and even leave the seals off. Then apart from an annual strip-down, just apply a drop of regular cycle oil once in a while. But don't expect it to actually feel any faster.

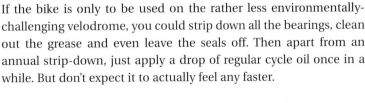

14. Don't Ring Us!

The main part of this chapter, like the one on suspension, was first published in 'Bike Culture Quarterly'. But unlike the suspension article, which is dating as I write, this section is, sad to say, timeless. For these are ideas that re-emerge year after year. By now you should understand what their failings are. But just in case you were not concentrating, I have included their shortcomings where appropriate.

B ent or curved cranks

One of the more enduring ideas that surfaces from time to time. The theory seems to be that by angling the crank forward, you can pedal through the 'dead' point more easily.

Wrong. There is no advantage in curving the crank, for all that matters is the distance between the pedal spindle and bottom bracket axle. You can form the materials that join them into the shape of a rams horn and it will make no difference apart from being heavy, weak and sheep-like.

O val/Biopace chainrings

Another regular, and one of the oldest. There is an example in the Science Museum in London from around 1890, arranged to give a big gear when pushing down (or horizontally on a recumbent) and a smaller one through the dead centre. This may not be such a bad idea, as some research has suggested a possible 10% increase in power output at lower than optimum pedalling rates. Unfortunately, other research shows that using regular pedals and cranks consumes all but 5% of the power we have in our little bodies.

Biopace on the other hand has never been shown to have an advantage other than by Shimano, who produced it. So why did everybody buy the stupid things? The main culprits must be the journalists who swallowed the hype and wrote what was expected of them. It actually did feel good, which was probably the intention, for having the gear get lower as you push down is very rewarding, and the bigger gear through the top dead centre is far less noticeable.

S olid tyres

A tyre that won't puncture seems like a jolly good idea. In fact, of all the bad things you can do to your bicycle, this is just about the worst.

They increase rolling resistance enormously, allow you and your bicycle to feel every bump and ripple in the road, have a much smaller contact area than pneumatics which, coupled with their hardness, means less grip. Their excessive weight gives any bike a leaden feel – and worst of all they don't puncture. The best thing about cycling is the variety it brings to your life, for you cannot enjoy the good days without bad ones to serve as a contrast. So when you

(Above) 1898: silly then, silly now. What matters is the distance between the pedal spindle and the bottom bracket axle.
(Below) Shimano's Biopace – 10/10 for marketing, 0/10 for function.

are struggling along into a headwind and driving rain, slowly realising that the occasional bump from the back is the valve bottoming out on the road, then you discover that you forgot to mend the spare tube... remember, tomorrow will be great.

Flywheels

Not something that reaches the shop often, but a favourite with inventors who reason that you could store up the energy released by whizzing downhill and use it to get you up the other side. The problem is that there isn't much spare energy to start with, and then with the losses incurred in converting motion into energy, storing it and then converting it back again, you end up with a small slice of a very small cake. Plus, of course, the real world is seldom a series of well defined ups and downs, so you are left pedalling a bicycle with ten kilos of expensive ballast on it.

One of the other myths about flywheels is that they carry you through hard spots. The very heavy disc wheels that Moser used in his first hour record ride were thought by many cyclists to have helped simply by acting as flywheels. Now when I was young and feckless and used to race motor cars, one of the first things that I did was to drill out the flywheel to reduce its weight, because any engineer will tell you that what a flywheel does is store energy, and storing energy is a bit like saving money in that it takes it out of circulation. Any energy that you have put into a flywheel is energy you have not put into moving your cycle forward.

Hand cranks

If I can go this fast using only my legs, how much faster could I go if I used my arms as well? Many people have asked this question, and some even built bikes and recently HPVs that allowed the rider to both steer and pedal with all limbs. I once thought it was quite a good idea, if a bit compli-

cated to arrange. I am indebted to my friend John Kingsbury of the Human Power Club for enlightening me.

It's all to do with the way we work. Our muscles are rather like electric motors and the heart and lungs like a battery supplying the power. The legs have the largest muscles in the body and can, especially after a bit of training, use all the power that the battery can deliver. Trying to run extra motors without any more power available simply means that you travel at the same speed on a heavy, complicated, expensive and ugly machine.

Auxiliary power

The search for auxiliary power has aroused a great deal of interest, for as you will have noticed there are many products which are promising easier cycling, from mini electric motors to bolt-on petrol engines.

Electric sounds the best idea because it is clean and quiet. Unfortunately battery technology is not up to it yet. They are heavy, expensive and full of things that are anything but clean, and of course the electricity still has to come from a power station. They can be made to work, but like the flywheel idea, you end up for most of the time pedalling a heavy, expensive bicycle.

More realistic is the bolt-on internal combustion engine. These are light, powerful and can be 'topped up' anywhere in the world and have all the power that you could need. The problem is that what you have created is a moped, and you can buy a purpose-built one for a lot less money.

Toothed belts and shaft drives

Getting rid of chains may be good for Marxists but it is a bad thing for cyclists. For the full technical explanation see the chapter on transmission. If you are a design student, this especially applies to you. What you should be designing is an easily-removable, non-rattling chain-guard.

Automatic gears

There have been several attempts to produce an 'automatic' bicycle over the years. Fortunately none of them reached the market.[16] But things may be about to change. In fact, by the time this book is available the latest attempt should be on the market.

Unlike most earlier designs that were based on the derailleur or expanding chainring, the newcomer uses an existing hub gear and is produced by Shimano. So if it is out there by 'now', it will work perfectly and sell by the thousand. But it still won't be what cyclists need.

This is because we are used to identifying 'automatic' with 'easy', and for us users of automatic cars, washing machines, etc. life is made easier. But for the motors that power these devices life is anything but easy. And without exception these automatics will use more energy than their manual equivalent.

So for the bicycle, where the power is supplied by one of the most reluctant engines available (but one with the most advanced engine management system available) automatics are the *last* thing we need. This is because we know *exactly* which gear to select. And at the end of the day if you can't be bothered to push the *button*, it is unlikely you will want to push on the *pedals*!

Variable length cranks

Hand cranks: very useful if you can't cycle with your legs. Otherwise not.

Another favourite with inventors. The idea is that, if you could arrange for the crank to be longer than normal on the power stroke, and shorter than normal on the return, you will get more power – but without the problems of leg stretch that a fixed length crank would cause.

Wrong again. You don't get more power with a long crank, sliding or otherwise. You just get a different gear. 'Power' is a function of force and speed combined. And for the devices that change the ratio of power to rest this reduces the muscle's rest time and so lowers its potential for generating thrust. The same effect can be produced by fitting oval chainrings the 'wrong' way, like Biopace, thus avoiding the need for a piece of mechanism that looks like part of a robot dishwasher.

As Richard Ballantine once said, with good bicycles the more you pay the less you get.

---

[16] *In the 1980s the 16-speed automatic Deal Drive bike got very close to production, order forms being distributed to potential customers. There have been a number of automatic gear systems for cycles, including the Fichtel & Sachs centrifugally controlled Automatic two-speed hub and Browning's microprocessor-controlled derailleur.*

Recommendations – Spending the Money!

A little tailpiece for those who were not concentrating – or who simply like turning to the back to see if there is a happy ending. Written in December '97 this advice will soon date. It should however give the reader an idea of how I think, as the following are what I would buy if I had to.

Time trialling and triathlon

For time trialling or triathlon, look for one of the LotusSport frames that may still be in production. Not actually designed by me and I get no royalties. That's another story – one my solicitor has advised me not to write.

Nonetheless LotusSport are probably the most aerodynamic you can buy. On the debit side, they are a little heavy and one or two have failed in use.

My own MCR by Giant is, I would guess, almost as aerodynamic, a lot more practical, and hopefully still in production.

Road racing

For road racing I have to start by mentioning my TCR. After all, that's what I designed it for, and if ONCE are using it, what can be wrong with it?

If you prefer a more traditional design, then the Trek OCLV is excellent – very light and stiff. If you are even more traditional, and can't give up the iron sticks, then a trip to Chas Roberts in Croydon is recommended. Cross his palm with silver and you could walk away with a work of art. Well, not actually walk away. More likely come back three months later – but it will still be very nice.

Velodrome

Velodrome use is a bit tricky. Mostly you will have to get the machine built to order by a small framebuilder, who won't like going by the advice I have given in this book!

For pursuit both the LotusSport and the MCR have bolt-on dropouts that could be replaced if you can get some replacements machined specially. But don't tell Giant that I told you.

Off-road

For off-road, there is no one design I would single out. There are lots of good composites and aluminium frames to choose from. Like everything else, you get what you pay for.

At least, that's what I wrote in the first draft. I'll revise that. There's no choice anymore for cross-country: Giant's new XTC.

Touring

For the tourist wanting something really good it's either the hand-built or maybe one of the Giant aluminium range – an Expedition or similar. You can easily change the handlebars, etc. to suit and it is a lot cheaper than starting from scratch.

Recumbents

As for recumbents I cannot recommend too highly the Windcheetah Speedy. This gem of a machine is without doubt the finest example of... Royalties? Oh yes, actually I do get some, so buy one for your friend as well.

If, like most, you prefer a two-wheeler, then the Kingcycle is an excellent sporty type, but is not currently in production. Or there is the Pashley PDQ. Based on the US Counterpoint Presto[17] design; this is less racy and more relaxed.

And finally...

And finally, don't buy a bicycle. Buy BICYCLES – lots of them, one for every day of the week, or even month. I am up to 20 so far.

It is wisely said that variety is the spice of life, so ride road and mountain, trade bike and folder, anything you can get your hands on. Store it in shed or garage, hang it on the wall or horde it in the loft.

Diamonds may be a girl's best friend, but the cycle in all its variety is mankind's.

---

[17] *Designed by Jim Weaver from whom Pashley purchased the design rights.*

Appendix

Reading Matter

Not the usual comprehensive list of every book in the world with 'bicycle' in the title, but a very personal view of some that I know of and merit a mention.

Publishers note: *In this section Mike Burrows has preferred to add updates to his original text, rather than change what he previously wrote. We respect his wishes.*

Periodicals

Magazines come and go. They change direction, get new editors, or fold entirely. So there are some 'last minute' updates to this list to reflect the changing market.

Cycling Weekly
IPC
Known popularly as 'the comic'. If you race a bike in the UK, you need *CW* if only to see that they have spelt your name correctly in the results. When I first started writing this book, the editor was Andy Sutcliffe. He was the best thing to happen to the mag in a long time. Andy brought the mountain bike into mainstream cycle racing and gave *CW*'s whole outlook a boost. Not that Andy and I always saw eye to eye, but it's a shame that he has moved on. His successors are 'competent' and *CW* is still the one to read if you race or follow racing.

Update: Sadly seems to be slipping backwards. Being true to its title 'Cycling' as opposed to 'Cycles' it reported the 1999 World Track Champs in Berlin extensively but without once mentioning bicycles,

and this was the last Worlds before the UCI ban on any cycle not designed in the 19th century came into force.

Mountain Bike Rider
IPC

Off-road mag from the *Cycling Weekly* office. As I write, it's the best of a rather disappointing bunch. It is at least aimed at an adult(ish) reader and well put together. But the road tests are about as accurate as the weather forecast.

Cycle Sport
IPC

Glossy monthly from the same team. Lots of pics of your favourite heroes riding up the big hills. Does what it does really well. But a shame about the road tests.

Mountain Biking UK
Future

The top selling UK bike mag for many years. It's the original wacky MTB magazine, claiming to target the 18-year-olds but actually reaching the 8-year-olds. If puberty is but a distant memory, you will hate it. If, on the other hand, you are still trying to get rid of the spots, this is the one for you.

That would normally be that, but *MBUK* is not like other magazines, because *MBUK* has on its inside back cover 'Mint Sauce'. In case you didn't know, this depicts the cartoon adventures of the world's greatest mountain biking sheep and his off-road pals. Not sure how the 8-year-olds cope with it; I would have thought you needed to be at least 27 and in line for a Turner prize to fully appreciate some of the issues. Oh yes, and it's laboriously scrawled by Jo Burt, whose doodles grace many of these pages. The same Jo Burt whom I can't really afford – hence the plug!

Total Mountain Bike
Future

Same team as *MBUK*. Lots of product pics and without doubt the shallowest and least useful road tests on the planet. A must for the young fashion victim.

Update: Gone! And not an issue too soon. Future recently launched a new mountain biking title called What Mountain Bike.

Cycling Plus
Future

Sensible all-round cycling mag – a rare thing these days. Probably the best road tests by a mainstream mag. Occasionally very interesting, but of necessity tends to repeat. Very good ongoing section on training, diet, fitness, etc.

Mountain Bike International
IPC

Was the best off-road mag but jumped on the free-ride/downhill bandwagon with a vengeance. It now appears to be trying to go straight again.

Update: There must be a lot of off-road on the road to Damascus, judging by how often this title changes direction. Now 100% race oriented and owned by IPC. Could slot in well with their current titles and team with this approach.

Pro Cycling
Cabal

New glossy road mag produced, but sadly not edited, by Andy Sutcliffe of *CW* fame. Interesting, lively format but especially bad road tests by 'old roadie'.

Maximum Mountain Bike
IPC

Just taken over by Andy Sutcliffe. Generally a good mag but in a very crowded market.

Update: Gone. Folded the same time as Total Bike.

Bike Culture Quarterly
Open Road

No advertising, no hype and no road tests, but lots of interesting stuff about all sorts of weird and wonderful machines that you wouldn't otherwise get to hear of. Produced by a well-meaning group of hippie bike nuts who need the money. Also available in German. Open Road, incidentally, are the company with the intelligence and foresight to publish this book.

Bycycle
Open Road

UK-focused and slightly more commercial (some adverts) bi-monthly from the hippies. Started off all pastel colours and preaching, but a new editor has recently taken over. So by the time you read this, it could well be the replacement for the old *New Cyclist*.

Update: Tragedy! Open Road is no more. the hippies have gone out of business. If business was not too strong a word: more like wealth re-distribution. However, not one but three phoenix have risen up. Peter Eland will launch Velo Vision magazine, Jim McGurn will stick to books (like this one), trading as the Company of Cyclists, and last I heard Dan Joyce was to be the editor of Cycletouring, where, hopefully, he will be able to nudge the sensible ones into the 21st Century.

Cycle Touring & Campaigning
Cyclists' Touring Club

The house magazine of the CTC. It's only available to members, which is no great loss to the world, except for its road tests. These are thorough, accurate and honest. They are carried out or supervised by Chris Juden, their excellent technical editor and a man who probably knows more about the bicycle than any other journalist.

Sadly, neither Chris nor any of his chums knows anything about publishing. As a result, what they produce is little more than a club newsletter on shiny paper. And they don't like mountain bikes or recumbents…

Update: Shock horror! The last issue of 1999 actually looked like a proper magazine. Not a great one, but a great leap forward for plus-four kind.

Also ran an excellent road test of a recumbent. There is hope for the chequered sock brigade.

Recumbent UK
New, obviously specialised mag for laid back types. It's well written in a bouncy, opinionated way. The road tests are okay but could be more critical.

A to B
A to B

Would you believe it? A little magazine dealing primarily with folding cycles and how to get them on trains, planes, etc. Lively and interesting, especially for those who keep their bikes in a bag!

Bicycling
Rodale (USA)

Was in its day the biggest and best – lots of tests and very techno. Its current claim to fame is probably the number of adverts for pick-up trucks. About to get a new editorial team as I write, so give it a look.

Update: New team seem to have put some substance back. Covers road and mountain, but lots of 'design' (usually a substitute for content), and little sense of authority.

Mountain Bike
Rodale (USA)

Originally a split-off from *Bicycling*. Now with the legendary Zapata Espinosa as editor, it does some of the best road tests around. Very fashion conscious, its restless layout uses lots of coloured type on coloured backgrounds. Nonetheless, still a good read.

Mountain Bike Action
Daisy/High Torque

The world's first ATB mag. Then with Zapata as editor, it set the standard for all other mags. Lots of action (always with a helmet, though)

and the ability to say something different about a dozen seemingly cloned ATBs. Good journalism but not very useful.

Bike Magazine
This American mag is like *National Geographic* with bikes. Beautiful photography and some nice rambling prose. Pick it up when you next get the chance.

Velo News
Inside Communications

Stars and stripes version of 'the comic', this is a fortnightly produced in the style of a large-format colour newspaper. Lots of race news and gossip about US and European road and dirt racing. A good read for the racer.

Fiets
This Dutch publication is probably the best general interest cycling magazine in the world. Sadly, its insistence on using the Dutch language throughout makes it impossible for me to be certain. Its editor, Guus van de Beek, has been at his desk for nearly 20 years, which must be a record. Buy a copy when you are next in Amsterdam.

Books

Richard's Bicycle Book
Richard Ballantine
Pan

The book that launched a thousand cyclists – well, many thousand probably. In all their forms, Richard's books have done more for cycling than anyone or anything has or could, short of turning the M25 into a velodrome. And as if that was not enough, as I write Richard is also putting pen to paper to produce a brand new, up-to-date version of the book. Intended as an introduction to cycling, you should not need it if you are reading this book. So buy it instead for a non-cycling friend.

Bicycling Science
Whitt & Wilson
MIT

If Richard's book was your starting point in the search for cycling's truths, and the volume you are now clutching is the next step, then *Bicycling Science* is the final destination. It's been the bible for cycle and HPV designers for many years. It covers every aspect of cycling in a scientific but understandable way. No humour, no pretty pics and (like all bibles) should be questioned occasionally. But a must if you are serious about the subject.

Bicycles and Tricycles
Archibald Sharp
MIT (reprint)

Already mentioned several times, this is a classic work. In effect, it's *Bicycling Science* but a hundred years earlier. It's even less funny and the science is not easy to understand – at least, not for cyclists of little brain like myself. But it's a must for enthusiasts.

The Sturmey-Archer Story
Tony Hadland
Pinkerton Press

A fascinating book by my co-conspirator and wordsmith. Fascinating, that is, for those who enjoy taking watches to pieces and then putting them back together. There is a lot more to the simple three-speed than a shiny body and a naff plastic shift lever!

Also from the same tireless author and publisher are the Moulton story, parts 1 & 2 – a must for fans of the small wheelers and a wonderful insight into Dr Moulton's single mindedness.

The Data Book
Van der Plas Publications

A wonderful collection of line drawings dating from the earliest cycles and turn of the century developments, but mostly by Rebour, the legendary illustrator for *Le Cycle*. It's cynically known as 'next year's Shimano catalogue', because of all the ideas shown that you will already know from their TLAs. (TLA is, of course, the three letter abbreviation for Three Letter Abbreviation.)

Bike Cult
David Perry

A bit of a mess really. An incredible amount of random information about all aspects of cycles and cycling, barely edited at all and with plenty of errors. But there's so much that you won't have seen elsewhere that it's worth adding to your collection.

On Your Bicycle
Jim McGurn
Open Road

Yes, that's right, the same Jim McGurn who is publishing this book, and a jolly nice chap he is, too. Not that he knows anything about bicycles as such, but that is not what his book is about. It is about 'cycling', which is of course what he and indeed most people are interested in. So, not a book that designers and engineers would normally choose to read… But they should, for it is about the consequences of what we did when we designed the bicycle in the first place. Something that we can be proud of, and which Jim puts down very well. A good read, even for those of us with broken fingernails.

Organisations

Union Cycliste Internationale
The governing body for cycle sport worldwide. In reality a bunch of **XXX ** ** ** ** XXX ***** ** XXXX ** *** **** XXXX** red hot steel ****** XXXX *** XXXX ** **** XXX * *** ** ****** small pieces **XXXX *** XXX **** ** XXXX ****** not even in a cardboard box.

British Cycling Federation
The UK's blazer brigade have had their problems but recently stood their ground against the UCI and their proposed 'rules to end cycling'. So a big thank-you for that. But I still get the feeling that they have yet to come to terms with the mountain bike. Well, it has only been around for 17 years!

Road Time Trials Council
Not exactly a bunch of radicals but at least they are not linked to the UCI. And they still control what is probably the most popular form of cycle sport in the country.

British Human Power Club
No blazers and no hang-ups, BHPC is for HPV enthusiasts everywhere. They run races, tours, etc and have a fun newsletter. Their booklet on 'how to do it' is an absolute must for the enthusiast.

Cyclists Touring Club
Not just for the checkered sock brigade. They were given a shake up when the Friends of the Earth started to do even more for 'real' cyclists than they did, and now do indeed work for all of us. I have been a member for 25 years and I have never worn silly socks.

Picture credits

Front cover: *Jason Patient*
Principal interior photography: *Sue Darlow, Jason Patient, Paul Batty, Paul Burrows, Richard Ballantine*
Illustrations and cartoons: *Jo Burt and Geoff Apps*
Charts and tables: *Whitt and Wilson's Bicycling Science (MIT Press, used with permission); Mike Burrows*

Front cover and pages 26, 88, 135, 141, 142: *Jason Patient.*
Pages 49, 57, 58, 59, 78, 83, 139: *Sue Darlow.*
Page 7 and page 82 (top): *Assignment Photography*
Page 8: *Tim Roy Photography*
All picture page 73: *Brian Donnan.*
Other pictures by *Paul Batty, Paul Burrows, Richard Ballantine.*
Printed by The Magazine Printing Company Enfield, Middx, EN3 7NT.